Problem Books in Mathematics

Edited by P.R. Halmos

Edward Lozansky Cecil Rousseau

Winning Solutions

Edward Lozansky
National Science Teacher's Association
Washington, DC 20009
USA

Cecil Rousseau
The University of Memphis
Memphis, TN 38152
USA

Series Editor

Paul R. Halmos
Department of Mathematics
Santa Clara University
Santa Clara, CA 95053
USA

Mathematics Subject Classification (1991): 11Axx 05Axx

Library of Congress Cataloging-in- Publication Data
Lozansky, Edward
Winning Solutions / Eward Lozansky, Cecil Roussseau
p.cm- (Problem books in mathematics)
Includes bibliographical references (p.-) and index.
ISBN 0-387-94743-4
1. Mathematics – problems,exercises, etc. I. Rousseau, Cecil II. Title.
III. Seriesw
QA43.L793 1996
510'.76-dc20

Springer Medizin Verlag
springer.de

Lozansky & Rousseau: Winning Solutions

First Indian Reprint 2010

ISBN 978-81-8489-526-1

This edition is manufactured in India for sale only in India, Pakistan, Bangladesh, Nepal and Sri Lanka and any other country as authorized by the publisher.

This edition is published by Springer (India) Private Limited, (a part of Springer Science+Business Media), Registered Office: 3rd Floor, Gandharva Mahavidyalaya, 212, Deen Dayal Upadhyaya Marg, New Delhi – 110 002, India

Printed in India by Rashtriya Printers, Delhi.

Preface

Problem-solving competitions for mathematically talented secondary school students have burgeoned in recent years. The number of countries taking part in the International Mathematical Olympiad (IMO) has increased dramatically. In the United States, potential IMO team members are identified through the USA Mathematical Olympiad (USAMO), and most other participating countries use a similar selection procedure. Thus the number of such competitions has grown, and this growth has been accompanied by increased public interest in the accomplishments of mathematically talented young people.

There is a significant gap between what most high school mathematics programs teach and what is expected of an IMO participant. This book is part of an effort to bridge that gap. It is written for students who have shown talent in mathematics but lack the background and experience necessary to solve olympiad-level problems. We try to provide some of that background and experience by pointing out useful theorems and techniques and by providing a suitable collection of examples and exercises.

This book covers only a fraction of the topics normally represented in competitions such as the USAMO and IMO. Another volume would be necessary to cover geometry, and there are other special topics that need to be studied as part of preparation for olympiad-level competitions. At the end of the book we provide a list of resources for further study.

A word of explanation is due the reader who is not already familiar with olympiads and the topics normally dealt with in such competitions. Until now, calculus has not been accepted as one of those topics. Problems on olympiad exams regularly call for use of Ceva's theorem, Chebyshev's inequality, the Chinese remainder theorem, and convex sets, *but not calculus*. The authors are the first to acknowledge that this book deals with an ecclectic list of topics. However, we have tried to choose these topics with the olympiad tradition and the needs of mathematically talented young persons in mind.

Many people have made valuable suggestions to us during the writing of this book. We are especially grateful to Basil Gordon (UCLA), Ian McGee (University of Waterloo), and Ron Scoins (University of Waterloo) for suggestions made concerning the first two chapters, and to David Dwiggins (University of Memphis) for his careful reading of the final manuscript.

The first two chapters of this book were written while one of the authors [CR] was on sabbatical at the University of Waterloo. This author wishes to thank Ron Dunkley for the invitation to visit Waterloo and to express his appreciation to all the members of the faculty and staff who helped make this visit a productive one.

For one of the authors [CR], the opportunity to write this book is an outgrowth of the good fortune of having been associated with both the USAMO and the IMO for many years. The opportunity for this author to play such a role was initially provided by Murray Klamkin, and has been supported and enlarged by many others, including Dick Gibbs, Samuel Greitzer, Walter Mientka, Ian Richards, Leo Schneider, many fine colleagues of the Mathematical Olympiad Summer Program (Titu Andreescu, Anne Hudson, Gregg Patruno, Gail Ratcliff, Daniel Ullman, Elizabeth Wilmer), and the many wonderfully talented students who have participated in the USAMO, IMO, and the Mathematical Olympiad Summer Program.

Finally, we are very grateful to the American Mathematical Competitions for permission to use problems from the AIME (American

Invitational Mathematics Examination) and the USAMO as examples and exercises in this book.

January, 1996

Edward Lozansky
Washington, D.C.

Cecil Rousseau
Memphis, TN

Contents

Preface v

1 Numbers **1**
1.1 The Natural Numbers 1
1.2 Mathematical Induction 11
1.3 Congruence 18
1.4 Rational and Irrational Numbers 29
1.5 Complex Numbers 35
1.6 Progressions and Sums 46
1.7 Diophantine Equations 56
1.8 Quadratic Reciprocity 65

2 Algebra **73**
2.1 Basic Theorems and Techniques 73
2.2 Polynomial Equations 92
2.3 Algebraic Equations and Inequalities 106
2.4 The Classical Inequalities 113

3 Combinatorics **141**
3.1 What is Combinatorics? 141
3.2 Basics of Counting 142

3.3 Recurrence Relations 149
3.4 Generating Functions 156
3.5 The Inclusion-Exclusion Principle 178
3.6 The Pigeonhole Principle 188
3.7 Combinatorial Averaging 195
3.8 Some Extremal Problems 202

Hints and Answers for Selected Exercises 215

General References 237

List of Symbols 239

Index 241

CHAPTER 1 Numbers

1.1 The Natural Numbers

Normally, we first learn about mathematics through **counting**, so the first set of numbers encountered is the set of counting numbers or **natural numbers** $\{1, 2, 3, \ldots\}$. Later, our knowledge is extended to **integers**, **rational numbers**, **real numbers** and **complex numbers**. A formal definition of even the natural number system requires careful thought, and one was given only in 1889 by the Italian mathematician Giuseppe Peano. Our approach is informal. It is assumed that the reader is familiar with various number systems. The following definitions ensure a common language with which to present problems and their solutions.

We use $\mathbb{Z}$ to denote the set of integers $\{\ldots, -2, -1, 0, 1, 2, \ldots\}$ and $\mathbb{Z}^+$ to signify the set of positive integers $\{1, 2, 3, \ldots\}$. We shall use the term **natural number** to mean a positive integer. (Mathematicians do not always agree on matters of terminology and notation. Some use the term natural number to mean a nonnegative integer.) If a and b are integers, we say that a **divides** b, in symbols $a|b$, if there is an integer c such that $b = ac$. Then a is a **divisor**, or **factor**, of b. A natural number $p > 1$ is said to be **prime** if 1 and p are its only positive divisors. A natural number $n > 1$ that is not prime is said to be **composite**. The list of prime numbers begins $2, 3, 5, 7, 11, 13, \ldots$ and has no end. The fact the list of primes is endless is proved in Euclid's *Elements*, and the proof given by Euclid is a model of the mathematician's art.

Theorem 1.1 (Euclid) *There are infinitely many primes.*

Euclid's argument is by contradiction. If $p_1, p_2, \ldots, p_N$ is a complete list of primes, then $M = (p_1 p_2 \cdots p_N) + 1$ is a number that is not divisible by any prime yet is larger than the largest prime, and this is impossible. In the first exercise at the end of this section, you have an opportunity to use the idea of Euclid's proof to show that the arithmetic progression $3, 7, 11, 15, 19, \ldots$ contains infinitely many primes.

If a and b are integers that are not both zero, their **greatest common divisor**, denoted by (a, b), is the largest natural number d such that $d|a$ and $d|b$. If $(a, b) = 1$, then a and b are said to be **relatively prime**. The following little fact is the cornerstone of a great theory.

Lemma 1.1 *Let a and b be two integers, not both zero. Then $\{xa + yb \mid x, y \in \mathbb{Z}\}$ is the set of all integral multiples of (a, b). In particular, (a, b) is the smallest positive number in this set.*

To prove this, let M be the set of all numbers of the form $xa + yb$ where $x, y \in \mathbb{Z}$ and note that if k is in M then any multiple of k is also in M, and if k and l are in M then $k - l$ is also in M. It is easy to see that M contains some positive integers, so M contains a *least* positive integer (see §1.2). Call this number d. Then M is just the set of all integral multiples of d. Otherwise, we could find a number $m \in M$ that when divided by d gives $m/d = q + r/d$ where $0 < r < d$, and this is impossible since $r = m - qd \in M$ and d is the least positive integer in M. Since $a, b \in M$ we have $d|a$ and $d|b$. On the other hand, since $d = x_0 a + y_0 b$ for some $x_0, y_0 \in \mathbb{Z}$, any number that divides both a and b also divides d. This clearly implies that $d = (a, b)$.

Using Lemma 1.1, we easily prove the following basic result.

Lemma 1.2 (Euclid's Lemma) *If $a|bc$ where a and b are relatively prime, then $a|c$.*

Since a and b are relatively prime, there are integers x_0 and y_0 such that $1 = x_0 a + y_0 b$ and thus $c = x_0 ac + y_0 bc$. Given that $a|bc$, we then have $a|c$.

Once we know that Euclid's Lemma is true, it is easy to use mathematical induction (§1.2) to prove the following important result.

Theorem 1.2 (Fundamental Theorem of Arithmetic) *Every natural number exceeding 1 is either prime or can be written uniquely (except for the order of factors) as the product of two or more primes.*

In symbols, the prime factorization result can be expressed as

$$n = \prod_p e^{e_p(n)}, \quad n > 1,$$

where $\prod_p$ denotes the product over primes and $e_p(n)$ is the **exponent** of p in the prime factorization of n. The important part of the theorem is the uniqueness of prime factorization. It is easy to see that every integer greater than 1 can be written as a product of primes. The part that isn't so obvious is that this can be done in only one way. But it is true, and the fact that prime factorization is unique provides a powerful method for solving many problems.

In view of their obvious importance, primes have been the object of many studies from the time of Euclid to the present, by both professional and amateur mathematicians. There are many **unsolved problems** concerning primes, problems that may look innocent enough but are (apparently) extremely difficult. For example, in a letter to Leonhard Euler written in 1742, Christian Goldbach conjectured that every even number greater than 2 is the sum of two primes. Goldbach's conjecture has never been proved or disproved. Of course we won't be trying to solve such problems. We limit our attention to problems that have easy solutions if one keeps the basic definitions and theorems in mind and views the question from the right perspective.

As an example of how the Fundamental Theorem can be applied, consider the following simple problem.

Example 1.1 *Find all primes p such that* $17p + 1$ *is a perfect square.*

Solution. Suppose $17p + 1 = n^2$. Then $17p = n^2 - 1 = (n-1)(n+1)$. The left-hand side expresses a certain number as the product of two primes. The right-hand side expresses the same number as the product of two factors. If either $n - 1$ or $n + 1$ were composite, then $(n - 1)(n + 1)$ would further factor and the final result would be the product of three or more primes. In view of the Fundamental Theorem, this can't be true. In fact, the uniqueness of prime factorization allows us to conclude that either $17 = n + 1$ and $p = n - 1$ or else $17 = n - 1$ and $p = n + 1$. The first alternative yields $p = 15$ and so doesn't work. The second gives $n = 18$ and $p = 19$. This does work, so $p = 19$ is the only prime for which $17p + 1$ is a perfect square. □

On occasion we will need to compute a greatest common divisor. An efficient way of doing this was found long ago by Euclid.

Theorem 1.3 (Euclidean Algorithm) *If $a > b > 0$ and a/b yields the quotient q and remainder r, in other words*

$$\frac{a}{b} = q + \frac{r}{b}, \qquad 0 \leq r < b,$$

then $(a, b) = (b, r)$.

To see why this statement is true, write the equation relating a, b, q, and r as $a = qb + r$, and note that any number that divides a and b also divides r and any number that divides b and r also divides a. Thus the set of common divisors of a and b is the same as the set of common divisors of b and r. In particular, the greatest common divisors are the same.

To calculate (a, b) using this theorem, we simply continue the process until the remainder is zero; thus we find the chain of equalities $(a, b) = (b, r) = \cdots = (d, 0) = d$. We *must* finally get a remainder that is zero since each remainder is nonnegative and the sequence of remainders is decreasing.

As an example, let us compute the greatest common divisor of 5999 and 994. The first division gives $5999/994 = 6 + 35/994$, the second gives $994/35 = 28 + 14/35$, the third gives $35/14 = 2 + 7/14$, and the last gives $14/7 = 2$ with remainder 0. We thus find $(5999, 994) = (994, 35) = (35, 14) = (14, 7) = (7, 0) = 7$.

Example 1.2 *Prove that if a and b are any two natural numbers, then*

$$(2^a - 1, 2^b - 1) = 2^{(a,b)} - 1.$$

Solution. Notice that with $a = qb + r$ $(0 \leq r < b)$, the following is an algebraic identity:

$$x^a - 1 = (x^b - 1)(x^{a-b} + x^{a-2b} + \cdots + x^{a-bq}) + x^r - 1.$$

In particular, setting $x = 2$ yields

$$2^a - 1 = (2^b - 1)Q + 2^r - 1,$$

where $Q = 2^{a-b} + 2^{a-2b} + \cdots + 2^{a-bq}$. Hence

$$(2^a - 1, 2^b - 1) = (2^b - 1, 2^r - 1).$$

Now, we realize that the chain of equalities in Euclid's algorithm $(a, b) = (b, r) = \cdots = (d, 0) = d$ is exactly matched by the chain $(2^a - 1, 2^b - 1) = (2^b - 1, 2^r - 1) = \cdots = (2^d - 1, 0) = 2^d - 1$. Thus, we have shown that $(2^a - 1, 2^b - 1) = 2^{(a,b)} - 1$ □

The following problem is quite simple, but it allows us to combine some of the things discussed so far.

Example 1.3 *Find a six-digit number that is increased by a factor of 6 if one exchanges (as a block) the first three digits of its decimal expansion with the last three.*

Solution. Write the desired number as $N = 1000A + B$, where A and B are three-digit numbers. Then exchanging (as blocks) the first three digits and the last three gives the number $1000B + A$. We are given that $1000B + A = 6(1000A + B)$ and it follows that $5999A = 994B$. Using $(5999, 994) = 7$, we find $857A = 142B$ and $(857, 142) = 1$. Now Euclid's Lemma tells us that $857|B$. But we also know that $0 < B < 10^3$. Hence $B = 857$, $A = 142$, and $N = 142857$.

Recall that n factorial is defined by

$$n! = n \cdot (n-1) \cdots 2 \cdot 1$$

for integers $n > 0$, and $0! = 1$. In the solution of various problems, it is helpful to know the prime factorization of $n!$. On the face of it, this seems like a hopelessly difficult task for all but the smallest values of n. For example, one might expect that a computer is required to find the prime factorization of a number as large as $20! = 2,432,902,008,176,640,000$. However, Adrien-Marie Legendre found a neat formula for the prime factorization of $n!$, and this result is helpful in solving many problems. To express Legendre's formula, we make use of the "floor," or "greatest integer," function. Specifically, $\lfloor x \rfloor$ denotes the greatest integer $\leq x$

Theorem 1.4 (Legendre) *The exponent of p in the prime factorization of n! is*

$$e_p(n!) = \sum_{r\geq 1} \left\lfloor \frac{n}{p^r} \right\rfloor .$$

This formula is fairly obvious once one thinks about it. Where do the factors of p come from in the factorization of $1 \cdot 2 \cdots n$? First of all, between 1 and n there are $\lfloor n/p \rfloor$ numbers that are multiples of p, namely $p,\ 2p, \ldots,\ \lfloor n/p \rfloor p$. These each contribute one factor of p. But of those, there are $\lfloor n/p^2 \rfloor$ multiples of p^2 and they each contribute *one more* factor of p. Now the $\lfloor n/p^3 \rfloor$ multiples of p^3 contribute one more factor still, and so it goes, until eventually $\lfloor n/p^r \rfloor = 0$.

Historical Note. The formula

$$n! = \prod_{p\leq n} p^{\lfloor n/p \rfloor + \lfloor n/p^2 \rfloor + \cdots}$$

is commonly called Legendre's formula, although it may have been discovered independently by various persons. It is in the second edition of Legendre's *Essai sur la théorie des nombres* (1808).

Using Legendre's formula, let us find the exponent of 5 in the prime factorization of 1000!. We get

$$\left\lfloor \frac{1000}{5} \right\rfloor + \left\lfloor \frac{1000}{5^2} \right\rfloor + \left\lfloor \frac{1000}{5^3} \right\rfloor + \left\lfloor \frac{1000}{5^4} \right\rfloor = 200 + 40 + 8 + 1 = 249.$$

This gives us an immediate solution to the following problem.

Example 1.4 *Find the number of terminal zeros in the decimal representation of* 1000!.

Solution. The answer is 249. The exponent of two in the prime factorization of 1000! is much larger than 249. (If you want to know what it is exactly, just use Legendre's formula.) Thus the highest power of ten that divides 1000! is the exponent of five in the prime factorization, namely 249. □

Example 1.5 *Show that if m and n are positive integers, then*

$$\frac{(2m)!(2n)!}{m!n!(m+n)!}$$

is an integer.

Solution. In view of Legendre's formula, it suffices to prove

$$\left\lfloor \frac{2m}{p^k} \right\rfloor + \left\lfloor \frac{2n}{p^k} \right\rfloor \geq \left\lfloor \frac{m}{p^k} \right\rfloor + \left\lfloor \frac{n}{p^k} \right\rfloor + \left\lfloor \frac{m+n}{p^k} \right\rfloor$$

for arbitrary p and k. More generally,

$$\lfloor 2a \rfloor + \lfloor 2b \rfloor \geq \lfloor a \rfloor + \lfloor b \rfloor + \lfloor a+b \rfloor,$$

holds for arbitrary real numbers a, b. To see this, observe that for any real number x we can write $x = \lfloor x \rfloor + \{x\}$ where $\{x\}$ denotes the **fractional part** of x and satisfies $0 \leq \{x\} < 1$. Making this substitution for both a and b above, the inequality reduces to $\lfloor 2x \rfloor + \lfloor 2y \rfloor \geq \lfloor x + y \rfloor$ where $x = \{a\}$ and $y = \{b\}$. Since $x + y < 2$, this inequality clearly holds if $x \geq 1/2$ or $y \geq 1/2$. If $x < 1/2$ and $y < 1/2$ both sides vanish, so the result holds in this case as well. □

The following problem, which was proposed for a recent International Mathematical Olympiad, gives us another good opportunity to use Legendre's formula. As background, recall that the **least common multiple** (LCM) of a collection of nonzero integers $\{a_1, a_2, \ldots, a_k\}$ is the smallest natural number that is divisible by each a_i. The **binomial coefficient** $\binom{n}{k}$ (read "*n choose k*") is defined by

$$\binom{n}{k} = \frac{n!}{k!(n-k)!}.$$

(For reference, $\binom{n}{k}$ is the number of ways to choose k elements from a set of size n, and it is the coefficient of $x^{n-k}y^k$ in the expansion of $(x+y)^n$.)

Example 1.6 *Prove that* $\binom{2n}{n}$ *divides* $LCM(1, 2, \ldots, 2n)$.

Solution. Let L denote the LCM of $\{1, 2, \ldots, 2n\}$ and let $M = \binom{2n}{n}$. To prove that $M|L$, it suffices to show that $e_p(M) \leq e_p(L)$ for each prime p. Since L is the smallest natural number that is divisible by each of $1, 2, \ldots, 2n$, it follows that $e_p(L)$ is the largest integer m such that $p^m \leq 2n$, namely $m = \lfloor \log(2n)/\log p \rfloor$. By Legendre's formula,

$$e_p(M) = e_p((2n)!) - 2e_p(n!) = \sum_{r=1}^{m} \left\{ \left\lfloor \frac{2n}{p^r} \right\rfloor - 2 \left\lfloor \frac{n}{p^r} \right\rfloor \right\}.$$

(Since $p^r > 2n$ for all $r > m$, we have included all of the nonvanishing terms in Legendre's formula.) Note that $\lfloor 2x \rfloor - 2\lfloor x \rfloor$ is either 0 or 1 (depending on whether or not $x - \lfloor x \rfloor$ is less than 1/2). In view of this fact, each of the terms in the above sum is either 0 or 1, and thus $e_p(M) \le m = e_p(L)$. □

The above result can be used in an elementary proof of a famous result of Chebyshev concerning primes. Let $\pi(x)$ denote the number of primes $\le x$. By elementary arguments, Chebyshev proved that there are positive constants A and B such that

$$A\frac{x}{\log x} < \pi(x) < B\frac{x}{\log x}.$$

(Here $\log x$ denotes the natural logarithm of x.) Later, Hadamard and de la Vallée Poussin independently proved that $\pi(x)\log(x)/x$ approaches 1 as x increases without bound. This is the famous **Prime Number Theorem**.

Example 1.7 *Let $\nu_p(n)$ denote the sum of the digits in the base p representation of n. Show that the exponent of p in the prime factorization of n! is related to $\nu_p(n)$ through*

$$e_p(n!) = \frac{n - \nu_p(n)}{p-1}.$$

Solution. Suppose $n = d_0 + d_1 p + \cdots + d_r p^r$ where $0 \le d_i < p$. Then

$$\left\lfloor \frac{n}{p} \right\rfloor = d_1 + d_2 p + \cdots + d_r p^{r-1},$$

$$\left\lfloor \frac{n}{p^2} \right\rfloor = d_2 + d_3 p + \cdots + d_r p^{r-2},$$

$$\vdots \quad \vdots \quad \vdots$$

$$\left\lfloor \frac{n}{p^r} \right\rfloor = d_r,$$

so by Legendre's formula

$$e_p(n!) = d_1 + d_2(p+1) + d_3(p^2+p+1) + \cdots + d_r(p^{r-1} + \cdots + 1)$$

and

$$(p-1)e_p(n!) = d_1(p-1) + d_2(p^2-1) + \cdots + d_r(p^r-1) = n - \nu_p(n). \quad □$$

In what follows, let $a(n) = e_2(n!)$ and let $b(n)$ denote the number of ones in the binary (base 2) expansion of n.

Example 1.8 *Let $B(m)$ denote the set of integers r for which 2^r is a term in the binary expansion of m. For example, $B(100) = \{2, 5, 6\}$. Prove that $\binom{n}{k}$ is odd if and only if $B(k) \subseteq B(n)$.*

Solution. The case $p = 2$ in the last problem yields $a(m) + b(m) = m$ for every nonnegative integer m. It follows that the exponent of two in the prime factorization of $\binom{n}{k}$ is

$$e_2\left(\binom{n}{k}\right) = a(n) - a(k) - a(n-k) = b(k) + b(n-k) - b(n).$$

Notice that $b(k) + b(n-k) - b(n) = 0$ if and only if there are no carries when k and $n-k$ are added in binary. It follows that $b(k) + b(b-k) - b(n) = 0$ if and only if $B(k) \subseteq B(n)$. From the relation given above, if $b(k) + b(n-k) - b(n) = 0$ then $\binom{n}{k}$ is odd; otherwise, $b(k) + b(n-k) - b(n) \geq 1$ and $\binom{n}{k}$ is even. We shall give a different proof of this result in §1.3. □

Our final example involving prime factorization introduces an important function in number theory.

Example 1.9 *If $n = p^r$ where p is prime, find the number of integers k where $1 \leq k < n$ and $(n, k) = 1$.*

Solution. Let us consider the numbers between 1 and $n = p^r$ that are *not* relatively prime to n. The only such numbers are the multiples of p, namely $p, 2p, 3p, \ldots, p^r$, and there are p^{r-1} such numbers. By deleting these numbers, we have left the numbers we want to count. Thus the number of integers k where $1 \leq k < p^r$ and $(p^r, k) = 1$ is

$$p^r - p^{r-1} = p^r\left(1 - \frac{1}{p}\right).$$

What about the case where n is not of the form p^r? For $n \geq 1$ let us define $\phi(n)$ to be the number of integers k where $1 \leq k \leq n$ and $(n, k) = 1$. It turns out that ϕ has a very nice property that allows us to compute its value in general by using the result of the last problem. It can be shown that if $(a, b) = 1$ then $\phi(ab) = \phi(a)\phi(b)$. (A function f with the property $f(ab) = f(a)f(b)$ whenever $(a, b) = 1$ is called **multiplicative**. Multiplicative functions play an important role in number theory.) It follows that if

$$n = p_1^{r_1} \cdots p_m^{r_m}$$

is the prime factorization of n, then

$$\phi(n) = p_1^{r_1}\left(1-\frac{1}{p_1}\right)\cdots p_m^{r_m}\left(1-\frac{1}{p_m}\right)$$
$$= n\left(1-\frac{1}{p_1}\right)\cdots\left(1-\frac{1}{p_m}\right).$$

This function is known as **Euler's ϕ-function** and it comes up again in §1.3. □

Exercises for Section 1.1

1. Prove that the arithmetic progression $3, 7, 11, 15, \ldots$ contains infinitely many primes. *Hint:* Suppose that $p_1, p_2, \ldots, p_N$ is a complete list of primes in this progression and consider the number $M = 4p_1p_2\cdots p_N - 1$. Of course, M is not divisible by 2, nor is it divisible by any of the primes $p_1, \ldots, p_N$. Is it possible that *every* prime divisor of M belongs to the progression $5, 9, 13, 17, 21\ldots$?
2. Find all natural numbers n for which $n^4 + 4$ is prime.
3. Find all natural numbers n such that $2^8 - 2^{11} + 2^n$ is a perfect square.
4. Find all four-digit numbers of the form $aabb$ that are perfect squares.
5. Prove that $(3n + 4, 2n + 3) = 1$ for every integer n.
6. Prove that the sum of the squares of five consecutive integers cannot be a perfect square.
7. Prove that the product of three consecutive natural numbers is never a perfect power (that is, a perfect square, perfect cube, or higher power). *Hint:* Write the product in question as $(n-1)n(n+1)$.
8. Suppose that N is a four-digit perfect square whose decimal digits are each less than 7, and that increasing each digit by 3 yields another perfect square. Determine N.
9. A number N is the sum of two squares if there exist integers a and b such that $N = a^2 + b^2$. Prove that if N is the sum of two squares, then so is $2N$. Generalize this result by proving that if M and N are each the sum of two squares, then so is MN.

10. In ancient Byzantium, a game similar to basketball was played that gave a points for a field goal and b points for a free throw. Surviving manuscripts do not record the values of a and b. However, it is known that in this game exactly 14 natural numbers were impossible point totals and that one of the impossible scores was 22. Determine the number of points given for field goals and free throws in Byzantine basketball. Assume that $a > b$. *Hint:* First prove the following general result. If a and b are relatively prime natural numbers, there are exactly $(a-1)(b-1)/2$ natural numbers that *cannot* be expressed in the form $xa + yb$ where x and y are nonnegative integers.
11. Prove that the product of any n consecutive integers is divisible by $n!$.
12. Find all n such that $2^{n-1} | n!$.
13. Find the number of terminal zeros in the decimal representation of $1 \cdot 4 \cdot 7 \cdots 1000$.
14. Find the prime factorization of 20!.
15. Prove that for $n > 1$ the nth **harmonic number**

$$H_n = 1 + \frac{1}{2} + \frac{1}{3} + \cdots + \frac{1}{n}$$

is not an integer.

1.2 Mathematical Induction

One often wishes to prove that a certain set of natural numbers is, in fact, the set of *all* natural numbers. This may be done through an appeal to the following important principle.

Principle 1.1 (Principle of Mathematical Induction) *Let T be a set of natural numbers that satisfies the following two conditions: (i) $1 \in T$, (ii) if n is an element of T, then $n + 1$ is also an element of T. Then $T = \mathbb{Z}^+$.*

The Principle of Mathematical Induction expresses a fundamental property of the natural numbers. In Peano's theory of the natural numbers, it is one of the axioms. A logically equivalent statement is the **Well-Ordering Principle**.

Principle 1.2 (Well-Ordering Principle) *Every nonempty subset of* $\mathbb{Z}^+$ *has a smallest element.*

In most applications of mathematical induction, we are given a sequence of statements $S_1, S_2, S_3, \ldots$, and T is taken to the set of all natural numbers n for which S_n is true. To prove that S_n is true for *every* natural number n, in other words $T = \mathbb{Z}^+$, we can appeal to the Principle of Mathematical Induction provided we can verify conditions (i) and (ii). Thus to prove that every statement in the sequence is true, we need to show (i) S_1 is true and (ii) S_{n+1} is true whenever its immediate predecessor S_n is true.

To prove the same thing using the Well-Ordering Principle, let F be the set of all natural numbers n for which S_n is false. To prove that F is empty, assume the contrary and consider its least element (which exists by the Well-Ordering Principle). The aim is then to show that this assumption leads to a contradiction.

As an illustration, let us use the Well-Ordering Principle to prove the Principle of Mathematical Induction. Thus we are given a set T that satisfies (i) and (ii) and we want to prove that $T = \mathbb{Z}^+$. Assume the contrary, namely assume that F (the complement of T in $\mathbb{Z}^+$) is nonempty, and let m be its smallest element. By (i) we know that $1 \in T$, so $m > 1$ and $m - 1$ is also a natural number. Since m is the *smallest* element of F, we conclude that $m - 1 \in T$. Thus our assumption that F is nonempty has forced us to conclude that there is a natural number m such that $m - 1 \in T$ and $m \notin T$. This contradicts (ii) so the assumption must be false. Hence $T = \mathbb{Z}^+$.

In Principle 1.1, (ii) may be replaced by the following: (ii)* if $1, 2, \ldots, n$ are elements of T, then $n + 1$ is also an element of T. We shall refer to this version as **strong induction**. Example 1.13 shows the use of induction in this form. Occasionally, it is appropriate to start the induction with $n = n_0 > 1$ rather than $n = 1$. Condition (i) in Principle 1.1 is then replaced by (i) T contains n_0, and the conclusion is that T is the set of all integers $n \geq n_0$. Example 1.14 uses this modification of the induction technique.

Many interesting problems can be solved by means of the Principle of Mathematical Induction. We begin by considering **sums**. The problem is to prove that the summation formula

$$\sum_{k=1}^{n} f(k) = F(n)$$

holds for every natural number n. Let T denote the set of natural numbers for which the formula is true. Then T satisfies condition (i) in the Principle of Mathematical Induction if $f(1) = F(1)$ and T satisfies condition (ii) if for every natural number n, the relation $f(1) + f(2) + \cdots + f(n) = F(n)$ implies $f(1) + f(2) + \cdots + f(n+1) = F(n+1)$. A moment's thought reveals that condition (ii) follows if $F(n) + f(n+1) = F(n+1)$. In other words, to prove by mathematical induction that our summation formula is true for every natural number n, we simply have to check that

(a) $f(1) = F(1)$,

(b) $F(n+1) - F(n) = f(n+1)$,

with the latter condition holding for arbitrary n. A simple illustration is in order.

Example 1.10 *Prove that for every natural number n,*

$$1 \cdot 2 + 2 \cdot 3 + \cdots + n(n+1) = \frac{n(n+1)(n+2)}{3}.$$

Solution. Set $f(n) = n(n+1)$ and $F(n) = n(n+1)(n+2)/3$. Now (a) is true since $f(1) = F(1) = 2$. To check (b), we evaluate $F(n+1) - F(n)$.

$$\begin{aligned} F(n+1) - F(n) &= \frac{(n+1)(n+2)(n+3)}{3} - \frac{n(n+1)(n+2)}{3} \\ &= (n+1)(n+2) \\ &= f(n+1). \end{aligned}$$

Thus (b) is true as well. □

There is a useful generalization of this result. Let m be a nonnegative integer. Then

$$\sum_{k=1}^{n} k(k+1)\cdots(k+m) = \frac{n(n+1)\cdots(n+m+1)}{m+2} \tag{1.1}$$

for every natural number n. For $n = 1$, both sides are equal to $(m+1)!$ so (a) holds. To check (b), we evaluate $F(n+1) - F(n)$, where $F(n)$ is the expression on the right side of (1.1). Simplifying the expression

$$\frac{(n+1)(n+2)\cdots(n+m+2)}{m+2} - \frac{n(n+1)\cdots(n+m+1)}{m+2},$$

we find

$$F(n+1) - F(n) = (n+1)(n+2)\cdots(n+m+1) = f(n+1).$$

Thus the formula holds in all cases. Using this formula, we can derive other sums. For example, since

$$k^3 = k(k+1)(k+2) - 3k(k+1) + k,$$

the sum of the first n cubes is given by

$$\frac{n(n+1)(n+2)(n+3)}{4} - 3\frac{n(n+1)(n+2)}{3} + \frac{n(n+1)}{2}.$$

Simplifying this expression, we obtain

$$\sum_{k=1}^{n} k^3 = \left[\frac{n(n+1)}{2}\right]^2.$$

Example 1.11 *Prove that $n(n-1)(n+1)(3n+2)$ is divisible by 24 for every natural number n.*

Solution. If $n = 1$ then $n(n-1)(n+1)(3n+2) = 0$, which is divisible by 24. Thus (i) of Principle 1.1 holds. To prove that (ii) holds, take $n \geq 1$ and assume that $24|n(n-1)(n+1)(3n+2)$. Then $24|(n+1)n(n+2)(3n+5)$ since

$$(n+1)n(n+2)(3n+5) = n(n-1)(n+2)(3n+2) + 12n(n+1)^2,$$

and $n(n+1)^2$ is always even. Thus our induction proof is complete. □

Problems involving **recurrence relations** are ideally suited for appoaches using mathematical induction. Here is a simple example.

Example 1.12 *Suppose that $a_1 = 3$ and $a_{n+1} = a_n(a_n + 2)$ for all $n \geq 1$. Find a general formula for the terms of this sequence.*

Solution. Computing the first three terms, we find $a_1 = 3 = 2^2 - 1$, $a_2 = 15 = 2^4 - 1$, $a_3 = 255 = 2^8 - 1$, and it is certainly tempting to conjecture that

$$a_n = 2^{2^n} - 1$$

for every natural number n. We have already verified that this is true for $n = 1$. Now take $n > 1$ and assume that the formula is true for the $(n-1)$st term of the sequence. Then

$$a_n = a_{n-1}(a_{n-1} + 2) = (2^{2^{n-1}} - 1)(2^{2^{n-1}} + 1) = 2^{2^n} - 1,$$

so our conjectured formula is correct. □

Example 1.13 *Suppose that $a_1, a_2, \ldots$ is a sequence of natural numbers that satisfies*

$$\sum_{k=1}^{n} a_k^3 = \left(\sum_{k=1}^{n} a_k\right)^2$$

for every natural number n. Is it necessarily true that $a_k = k$ for each k?

Solution. The answer is "yes." First of all, for $n = 1$ the given condition yields $a_1^3 = a_1^2$. Since $a_1 \neq 0$, we have $a_1 = 1$. Now assume that $a_k = k$ for $k = 1, 2, \ldots, n$ (strong induction) and ask whether or not $a_{n+1} = n + 1$. From what we are given,

$$1^3 + \cdots + n^3 + a_{n+1}^3 = (1 + \cdots + n + a_{n+1})^2,$$

and it follows that

$$\left[\frac{n(n+1)}{2}\right]^2 + a_{n+1}^3 = \left[\frac{n(n+1)}{2}\right]^2 + n(n+1)a_{n+1} + a_{n+1}^2.$$

Simplifying, we get

$$(a_{n+1} - (n+1))(a_{n+1} + n) = 0,$$

and (since a_{n+1} is positive) it follows that $a_{n+1} = n + 1$. Thus $a_k = k$ for every natural number k follows by induction. □

Our final example of a proof using mathematical induction is actually a special case of a famous theorem proved by Paul Turán in 1941. The problem as stated was given on the 1956 William Lowell Putnam Competition.

Example 1.14 (1956 Putnam) *Consider a set of 2n points in space, $n > 1$. Suppose they are joined by at least $n^2 + 1$ segments. Show that at least one triangle is formed. Show that for each n it is possible to have 2n points joined by n^2 segments without a triangle being formed.*

Solution. Consider first the case $n = 2$. Between four points there are in all six possible segments. Hence, if the four points are joined by five segments, only one segment is missing. If AB is the missing segment, then any three of the points not including both A and B form a triangle. Thus the statement is true for $n = 2$.

Now we use mathematical induction. Take $n > 2$ and assume that if $2(n-1)$ points are joined by at least $(n-1)^2 + 1$ segments, then at least one triangle is formed. We are now given $2n$ points joined by at least $n^2 + 1$ segments and we want to prove that there is a triangle. Consider two points that are joined by a segment. Call these points P and Q and note that if any of the remaining $2(n-1)$ points are joined to *both* P and Q, a triangle is formed.

If this is not the case, then the number of segments between $\{P, Q\}$ and the remaining $2(n-1)$ points is at most $2(n-1)$ since each of the remaining points may be joined to either P or Q but not both. Delete P, Q, and all incident segments. We are left with $2(n-1)$ points joined by at least $(n^2+1)-2(n-1)-1 = (n-1)^2+1$ segments. Thus a triangle is formed.

If the $2n$ points are divided into two disjoint sets of n points apiece and only the n^2 segments joining points in one set with points in the other are used, then no triangle is formed. □

Exercises for Section 1.2

1. For each of the following, prove that the given statement is true for every natural number n.

   (a) $1^2 + 3^2 + \cdots + (2n-1)^2 = \frac{n(4n^2-1)}{3}$.

   (b) $1 \cdot 1! + 2 \cdot 2! + \cdots + n \cdot n! = (n+1)! - 1$.

   (c) $\left(1 - \frac{1}{4}\right)\left(1 - \frac{1}{9}\right)\cdots\left(1 - \frac{1}{(n+1)^2}\right) = \frac{(n+2)}{2(n+1)}$.

   (d) $6 \mid n(n^2+5)$.

2. Show that

$$\sum_{k=1}^{n} \frac{1}{k(k+1)(k+2)} = \frac{1}{2}\left[\frac{1}{2} - \frac{1}{(n+1)(n+2)}\right]$$

   using mathematical induction.

3. Suppose that $x_1 = 2$, $x_2 = 3$, and $x_{n+2} = 3x_{n+1} - 2x_n$ for all $n \geq 1$. By looking at the first few terms, guess a general formula for x_n. Prove that the formula holds for every natural number n using mathematical induction.

4. Observe that

$$1^2 = 1 \cdot 2 \cdot 3/6,$$
$$1^2 + 3^2 = 3 \cdot 4 \cdot 5/6,$$
$$1^2 + 3^2 + 5^2 = 5 \cdot 6 \cdot 7/6.$$

Based on these examples, guess a general formula. Prove it using mathematical induction.

5. Derive a formula for the sum $1^4 + 2^4 + \cdots + n^4$.

6. Prove that for every real number $x > -1$ and every natural number n,

$$(1 + x)^n \geq 1 + nx.$$

This is **Bernoulli's inequality**.

7. If x and y are any two real numbers, then $|x + y| \leq |x| + |y|$. Using this fact, prove that if $x_1, x_2, \ldots, x_n$ are any n real numbers, then

$$|x_1 + x_2 + \cdots x_n| \leq |x_1| + |x_2| + \cdots + |x_n|.$$

8. Suppose that $0 \leq x_i \leq 1$ for $i = 1, 2, \ldots, n$. Prove that

$$2^{n-1}(1 + x_1 x_2 \cdots x_n) \geq (1 + x_1)(1 + x_2) \cdots (1 + x_n),$$

with equality if and only if $n - 1$ of the x_i's are equal to 1.

9. The sequence of integers $F_1, F_2, F_3, \ldots$ defined by $F_1 = 1$, $F_2 = 1$, and $F_n = F_{n-1} + F_{n-2}$ for $n \geq 3$ is called the **Fibonacci sequence**. Prove that for every natural number n, the Fibonacci numbers F_n and F_{n+1} are relatively prime.

10. Prove that for every natural number n,

$$F_n = \frac{a^n - b^n}{\sqrt{5}}$$

where $a = \frac{1+\sqrt{5}}{2}$ and $b = \frac{1-\sqrt{5}}{2}$.

11. Prove that

$$F_1^2 + F_2^2 + \cdots + F_n^2 = F_n F_{n+1}.$$

12. Prove that for every natural number n,

$$\left\lfloor \left(\frac{7 + \sqrt{37}}{2} \right)^n \right\rfloor$$

is divisible by 3. [Power Question, 1978 New York State Mathematics League] *Hint:* For $n = 1, 2, 3, \ldots$ let

$$a_n = \left(\frac{7 + \sqrt{37}}{2}\right)^n + \left(\frac{7 - \sqrt{37}}{2}\right)^n.$$

Use mathematical induction to prove that $a_{n+2} = 7a_{n+1} - 3a_n$ and that $a_n - 1$ is divisible by 3 for every natural number n. Prove that

$$a_n - 1 = \left\lfloor \left(\frac{7 + \sqrt{37}}{2}\right)^n \right\rfloor.$$

13. The vertices of a convex polygon with $2n + 1$ sides are colored so that no two consecutive vertices have the same color. Prove that the polygon can be divided into triangles using only nonintersecting diagonals having endpoints of different color. (A **convex region** has the property that any segment joining two of its points lies entirely within the region.) [1978 Jóseph Kűrschák Competition]

14. Given a $(2m + 1) \times (2n + 1)$ checkerboard where the four corner squares are black, show that if one removes any one red and two black squares, the remaining board can be covered with dominoes (1×2 rectangles).

15. Suppose that natural numbers $x_1, x_2, \ldots, x_n$ are such that $x_1 | (x_n + x_2)$, $x_2 | (x_1 + x_3), \ldots, x_n | (x_{n-1} + x_1)$. (Think of the n numbers as being on a circle, so $x_{n+k} = x_k$. Then each number divides the sum of its two neighbors.) Prove that

$$\frac{x_n + x_2}{x_1} + \frac{x_1 + x_3}{x_2} + \cdots + \frac{x_{n-1} + x_1}{x_n} \leq 3n - 1.$$

1.3 Congruence

Sometimes it takes a great mind to think of a simple idea. An example is the invention of the simple but powerful idea of congruence by the great German mathematician Carl Friedrich Gauss. Let a and b

be integers and let m be a natural number. Following Gauss, let us write

$$a \equiv b \pmod{m} \qquad \text{if} \qquad m \mid (a-b).$$

The statement $a \equiv b \pmod{m}$ is read ***a* is congruent to *b* modulo *m***; the number m is called the **modulus** of the congruence. Congruence modulo m divides the set of all integers into m subsets called **residue classes**. (Example: for $m = 2$, the two classes are the *even integers* and the *odd integers*.) Integers a and b are in the same class if $a \equiv b \pmod{m}$ and they are in different classes if $a \not\equiv b \pmod{m}$. Statements involving congruence are easily manipulated using the following basic properties. Each of these properties is easily proved from the definition of congruence.

Property 1.1 *The congruence $a \equiv a \pmod{m}$ holds for every integer a; congruence is reflexive.*

Property 1.2 *If $a \equiv b \pmod{m}$ then $b \equiv a \pmod{m}$; congruence is symmetric.*

Property 1.3 *If $a \equiv b \pmod{m}$ and $b \equiv c \pmod{m}$, then $a \equiv c \pmod{m}$; congruence is transitive.*

Property 1.4 *If $a \equiv b \pmod{m}$ and $c \equiv d \pmod{m}$, then $a + c \equiv b + d \pmod{m}$; congruences may be added.*

Property 1.5 *If $a \equiv b \pmod{m}$ and $c \equiv d \pmod{m}$, then $ac \equiv bd \pmod{m}$; congruences may be multiplied.*

Property 1.6 *If $ab \equiv ac \pmod{m}$ and $(a, m) = 1$, then $b \equiv c \pmod{m}$; in other words, we may divide both sides of a congruence by a number that is relatively prime to m.*

For practice using the concept of congruence let's prove Property 1.6. We are given that $ab \equiv ac \pmod{m}$. Thus we know that $m \mid a(b - c)$. But we are also given $(a, m) = 1$. It follows from Euclid's Lemma that $m \mid (b - c)$, in other words, $b \equiv c \pmod{m}$.

Let p be a prime and let $\mathbb{Z}_p$ denote the set $\{0, 1, 2, \ldots, p - 1\}$. Define addition $(+)$ and multiplication $(\cdot)$ on $\mathbb{Z}_p$ using congruence modulo p. Thus $a + b = c$ for $a, b, c \in \mathbb{Z}_p$ if the ordinary sum of a and b is congruent to c modulo p. Likewise, $a \cdot b = c$ if the ordinary product of a and b is conguent to c modulo p. Using the properties of congruence, it is easy to verify that the system $(\mathbb{Z}_p, +, \cdot)$ enjoys the customary features of arithmetic. Addition and multiplication are commutative and associative, the distributive law holds, there is an identity element for addition (0), and there is an identity element for multiplication (1). For each $a \in \mathbb{Z}_p$ there is an element $-a$ such that $(-a) + a = 0$. Finally, for each nonzero element $a \in \mathbb{Z}_p$ there is an element $a^{-1} \in \mathbb{Z}_p$ such that $a^{-1} \cdot a = 1$. This follows by using Euclid's Lemma and the fact that p is prime. The system $(\mathbb{Z}_p, +, \cdot)$ is an example of a **finite field**.

The way in which congruence arithmetic can effectively simplify problems involving divisibility is illustrated by the following example.

Example 1.15 *Prove that* $36^{36} + 41^{41}$ *is divisible by* 77.

Solution. Since $36 \equiv 1 \pmod 7$ and

$$36^5 \equiv 3^5 \equiv 1 \pmod{11},$$

we find that

$$36^5 \equiv 1 \pmod{77}.$$

This lets us cut the problem down to size. Since $41 \equiv -36 \pmod{77}$, we have

$$\begin{aligned}(36)^{36} + 41^{41} &\equiv (36)^{36} + (-36)^{41} \\ &\equiv (36)^{36}\left[1 - (36)^5\right] \\ &\equiv 0 \pmod{77}. \quad \square\end{aligned}$$

As this problem illustrates, a calculation modulo m involving a power a^k where k is large may be simplified considerably if we can find an exponent r such that $a^r \equiv 1 \pmod m$. For the case of a prime modulus p, the following theorem of Fermat shows that $p-1$ is such an exponent.

Theorem 1.5 (Fermat's Little Theorem) *If p is prime and a is not divisible by p, then*

$$a^{p-1} \equiv 1 \pmod p.$$

Fermat's Theorem was generalized by Euler to the case when the modulus isn't necessarily prime. Recall from §1.1 that the Euler function is given by

$$\phi(m) = m \prod_{p|m} \left(1 - \frac{1}{p}\right).$$

Theorem 1.6 (Euler's Theorem) *If $(a, m) = 1$, then*

$$a^{\phi(m)} \equiv 1 \ (mod\ m),$$

where ϕ is the Euler function.

Here is a nice proof of Euler's Theorem due to James Ivory. Set $k = \phi(m)$ and let $r_1, r_2, \ldots, r_k$ be a listing of the positive integers $r \le m$ such that $(r, m) = 1$. Consider the sequence $r_1a, r_2a, \ldots, r_ka$. We claim that modulo m the numbers in this sequence form a permutation of $r_1, r_2, \ldots, r_k$. Indeed, since $(r_ia, m) = 1$, the residue class containing r_ia must be represented by some r_j and $r_ia \equiv r_k \pmod{m}$ implies $r_i \equiv r_k \pmod{m}$ by Property 1.6. Thus

$$(r_1a) \cdot (r_2a) \cdots (r_ka) \equiv r_1 \cdot r_2 \cdots r_k \pmod{m},$$

from which we obtain $a^k \equiv 1 \pmod{m}$ by applying Property 1.6.

Example 1.16 *Find the last three digits of 7^{9999}.*

Solution. Since $\phi(1000) = 1000(\frac{1}{2})(\frac{4}{5}) = 400$, we know from Euler's Theorem that

$$7^{10000} = (7^{400})^{25} \equiv 1 \pmod{1000}.$$

At this point, it is useful to note that $7 \cdot 143 = 1001$ so $7 \cdot 143 \equiv 1 \pmod{1000}$. Using basic properties of congruence, we find

$$7^{9999} \equiv 143 \cdot 7 \cdot 7^{9999} \equiv 143 \cdot 7^{10000} \equiv 143 \pmod{1000}.$$

Thus the decimal representation of 7^{9999} ends with 143. □

The following two problems are more challenging. The solutions show the power of combining some of the problem-solving methods that we have discussed so far.

Example 1.17 (1991 USAMO) *Show that for any fixed integer $n \ge 1$ the sequence*

$$2,\ 2^2,\ 2^{2^2},\ 2^{2^{2^2}}, \ldots \ (mod\ n)$$

is eventually constant.

Solution. Our proof is by induction on n. The case $n = 1$ is clear. Now take $n > 1$ and suppose that the result is true for all positive integers less than n. We distinguish two cases.

Case 1: n is even. Write $n = 2^k q$ where $k \geq 1$ and q is odd. By induction, the sequence $a_1, a_2, a_3, \ldots$ is eventually constant modulo q. Clearly $a_i \equiv 0 \pmod{2^k}$ for all sufficiently large i. Since 2^k and q are relatively prime, $2^k|(a_{i+1}-a_i)$ and $q|(a_{i+1}-a_i)$ imply $n|(a_{i+1}-a_i)$. Thus the sequence $a_1, a_2, a_3, \ldots$ is eventually constant modulo n.

Case 2: n is odd. In this case, there exists a positive integer $r < n$ such that

$$2^r \equiv 1 \pmod{n}. \tag{1.2}$$

Indeed, Euler's Theorem yields (1.2) with $r = \phi(n)$. By induction, the sequence $a_1, a_2, a_3, \ldots$ is eventually constant modulo r. But in view of (1.2), $a_i \equiv c \pmod{r}$ implies

$$a_{i+1} = 2^{a_i} = 2^{m_i r + c} = (2^r)^{m_i} 2^c \equiv 2^c \pmod{n}.$$

Thus $a_1, a_2, a_3, \ldots$ is eventually constant modulo n. □

Example 1.18 (1972 Putnam) *Prove that there is no integer $n > 1$ for which $n|(2^n - 1)$.*

Solution. We use the Well-Ordering Principle. Suppose that the set

$$S = \{n \mid n > 1,\ n|(2^n - 1)\}$$

is nonempty and let m be its smallest element. Clearly m must be odd; thus $m|(2^{\phi(m)}-1)$ by Euler's Theorem. Let $d = (m, \phi(m))$. Since $m|(2^m - 1)$ and $m|(2^{\phi(m)} - 1)$, it follows from the result in Example 1.2 that $m|(2^d - 1)$. Since $m > 1$ and $m|(2^d - 1)$, we know that $d > 1$. Also $d|(2^d - 1)$ because $d|m$ and $m|(2^d - 1)$. Since $d \leq \phi(m) < m$ we have reached a contradiction by producing an element that belongs to S and is smaller than m. This contradiction shows that our original assumption is false and S is empty. □

Since $a - b$ divides itself and

$$a^k - b^k = (a-b)(a^{k-1} + a^{k-2}b + \cdots + b^{k-1})$$

for $k \geq 2$, we see that $a^k - b^k$ is divisible by $a - b$ for all $k \in \mathbb{Z}^+$. This observation leads to the following useful fact.

Theorem 1.7 *Let P be a polynomial with integral coefficients, and let a and b be arbitrary integers. Then $P(a) - P(b)$ is divisible by $a - b$. In other words, $P(a) \equiv P(b) \pmod{a-b}$.*

The following well-known fact comes about in exactly the same way.

Theorem 1.8 *Let $s(n)$ denote the sum of the digits in the decimal representation of n. Then $n \equiv s(n) \pmod 9$.*

Example 1.19 (1974 USAMO) *Let a, b, and c denote three distinct integers, and let P denote a polynomial having all integral coefficients. Show that it is impossible that $P(a) = b$, $P(b) = c$, and $P(c) = a$.*

Solution. Suppose $P(a) = b$, $P(b) = c$, and $P(c) = a$ for distinct $a, b, c \in \mathbb{Z}$. Then $a - b$ divides $P(a) - P(b) = b - c$. In like manner, $b - c$ divides $c - a$ and $c - a$ divides $a - b$. Thus our assumption leads to $|a - b| = |b - c| = |c - a|$, which is impossible for distinct integers a, b, c. □

Now we look at **simultaneous congruences** and the **Chinese Remainder Theorem**. First let us point out the following simple fact.

Lemma 1.3 *Suppose that $(a, b) = 1$. Then there exists an integer x such that $ax \equiv 1 \pmod b$.*

Indeed, according to Lemma 1 there are integers x and y such that $xa + yb = (a, b) = 1$, and this yields the desired x. We use this result in our proof of the following important theorem.

Theorem 1.9 (Chinese Remainder Theorem) *Suppose that $m_1, m_2, \ldots, m_k$ are pairwise relatively prime and $a_1, a_2, \ldots, a_k$ are arbitrary integers. Then there exist solutions of the simultaneous congruences*

$$x \equiv a_i \pmod{m_i}, \qquad i = 1, 2, \ldots, k. \tag{1.3}$$

Any two solutions are congruent modulo M, where

$$M = m_1 m_2 \cdots m_k.$$

To see why this is true, let us first note that the problem of solving the given system can be reduced to that of solving a special sequence of congruences. Suppose that for each i, we can find a number x_i such that

$$x_i \equiv \begin{cases} 1 \pmod{m_j}, & j = i, \\ 0 \pmod{m_j}, & j \neq i. \end{cases}$$

Then $x = a_1x_1 + a_2x_2 + \cdots + a_kx_k$ gives the desired solution. To obtain the required x_i, we proceed as follows. To ensure that $x_i \equiv 0 \pmod{m_j}$ for $j \neq i$, set $x_i = (M/m_i)z_i$. Then to satisfy $x_i \equiv 1 \pmod{m_i}$, we need to find z_i such that $(M/m_i)z_i \equiv 1 \pmod{m_i}$. Since $(m_i, M/m_i) = 1$, the existence of z_i follows from our lemma.

Let x and y be any two solutions of (1.3). Then $x - y$ is divisible by m_i for $i = 1, 2, \ldots, k$ and so $x - y$ is divisible by M. In other words, any two solutions are congruent modulo M.

Here is a simple application of the Chinese Remainder Theorem. A **square-free** integer has no repeated prime factor; for example, $15 = 3 \cdot 5$ is square-free and $12 = 2^2 \cdot 3$ is not.

Example 1.20 *Prove that for every n there is a sequence of n consecutive natural numbers, none of which is square-free.*

Solution. Let $p_1, p_2, \ldots, p_n$ denote the first n primes. By the Chinese Remainder Theorem, we can find x such that

$$x \equiv -i \pmod{p_i^2}, \qquad i = 1, 2, \ldots, n.$$

Now each number in the sequence $x + 1,\ x + 2,\ \ldots,\ x + n$ has a repeated prime factor. In fact, $x + i$ is divisible by p_i^2. Thus we have found n consecutive integers, none of which is square-free. □

Example 1.21 (1986 USAMO) *(a) Do there exist 14 consecutive positive integers each of which is divisible by one or more primes p from the interval $2 \le p \le 11$? (b) Do there exist 21 consecutive positive integers each of which is divisible by one or more primes p from the interval $2 \le p \le 13$?*

Solution (a) No. Suppose such integers exist. On the list of 14 consecutive integers, the 7 that are even certainly satisfy the condition. Thus we can focus on the odd integers. Let these be $a, a + 2, a + 4, a + 6, a + 8, a + 10, a + 12$. At most three of these numbers are divisible by 3, at most two are divisible by 5, at most one is divisible by 7, and at most one is divisible by 11. If each of these is divisible by one of the primes 3, 5, 7, 11, there is no room to spare; three of the numbers are divisible by 3, two are divisible by 5, one is divisible by 7, one is divisible by 11, and none is divisible by more than one of the primes 3, 5, 7, 11. The numbers divisible by 3 must be a, $a + 6$ and $a + 12$, and two of the remaining numbers $a + 2, a + 4, a + 8, a + 10$ must be divisible by 5. This is clearly impossible.

(b) Yes. If n is divisible by $210 = 2 \cdot 3 \cdot 5 \cdot 7$ then with the possible exception of $n-1$ and $n+1$, every term in the sequence of 21 consecutive integers $n-10, n-9, \ldots, n+9, n+10$ is divisible by one of the primes 2, 3, 5, 7. Thus we have the desired sequence of positive integers if $n-1$ is divisible by 11 and $n+1$ is divisible by 13, or vice versa. Altogether, our needs are met if n satisfies

$$n \equiv 0 \pmod{210}$$
$$n \equiv 1 \pmod{11}$$
$$n \equiv -1 \pmod{13}.$$

For completeness, we determine the solution of this system. Note that $n = 210$ satisfies the first two congruences, and thus so does $n = 2310k + 210$ for any k. To satisfy the last congruence, we choose k so that $2310k \equiv -211 \pmod{13}$. This simplifies to $9k \equiv -3 \pmod{13}$ and is clearly satisfied by $k = 4$. Thus $n = 4 \cdot 2310 + 210 = 9450$ is a solution (in fact, the smallest positive solution) and $n = 9450 + 30030m$ is the general solution. □

Example 1.22 *Prove that there is a positive integer k such that $k \cdot 2^n + 1$ is composite for every nonnegative integer n.*

Solution. Consider the numbers in the table below.

i	1	2	3	3	5	6
a_i	0	0	1	3	7	23
b_i	2	3	4	8	12	24
p_i	3	7	5	17	13	241

It is easily checked that these numbers satisfy the following conditions.

- $p_1, p_2, \ldots, p_6$ are distinct primes,
- $0 \le a_i < b_i$ and $p_i | (2^{b_i} - 1)$ for $1 \le i \le 6$,
- Each nonnegative integer n satisfies $n \equiv a_i \pmod{b_i}$ for at least one $i \le 6$. [It suffices to check that this is the case for $n = 0, 1, 2, \ldots, 23$.]

By the Chinese Remainder Theorem, we can choose k so that

$$k \equiv -2^{b_i - a_i} \pmod{p_i}, \quad i = 1, 2, \ldots, 6.$$

Specifically, $k = 1624097$ satisfies all six congruences. Then for every nonnegative integer n, there are integers q and j such that $n = b_j q + a_j$. Since $k \equiv -2^{b_j - a_j} \pmod{p_j}$ and $2^{b_j} \equiv 1 \pmod{p_j}$ we have

$$k \cdot 2^n + 1 \equiv -2^{b_j - a_j} 2^{q b_j + a_j} + 1 \equiv 0 \pmod{p_j}.$$

Since k is greater than each of the primes 3, 5, 7, 13, 17, 241, it is clear that $k \cdot 2^n + 1$ is composite for each $n \geq 0$. □

Now we present the alternative proof of the result in Example 1.8, as promised earlier. Write $P(x) \equiv Q(x) \pmod 2$ to mean that corresponding coefficients of polynomials P and Q are congruent modulo 2. Then

$$(1 + x)^2 = 1 + 2x + x^2 \equiv 1 + x^2 \pmod 2,$$

and more generally

$$(1 + x)^{2n} = [(1 + x)^2]^n \equiv (1 + x^2)^n \pmod 2.$$

Thus by induction

$$(1 + x)^{2^r} \equiv 1 + x^{2^r} \pmod 2, \qquad r \geq 0. \tag{1.4}$$

To work out the binomial expansion of $(1 + x)^n \pmod 2$, we can use the binary expansion of n together with (1.4). Thus, for example, $(1 + x)^{10} = (1 + x)^8(1 + x)^2 \equiv (1 + x^8)(1 + x^2) \equiv (1 + x^2 + x^8 + x^{10}) \pmod 2$, from which it follows that $\binom{10}{0}$, $\binom{10}{2}$, $\binom{10}{8}$, and $\binom{10}{10}$ are odd and the remaining binomial coefficients in the expasion of $(1 + x)^{10}$ are even. Now to the general case. As before, let

$$B(m) = \{r \mid 2^r \text{ is a term in the binary expansion of } m\}.$$

Then

$$\begin{aligned} \sum_{k=0}^{n} \binom{n}{k} x^k &= (1 + x)^n \\ &= \prod_{r \in B(n)} (1 + x)^{2^r} \end{aligned}$$

$$\equiv \prod_{r \in B(n)} (1 + x^{2^r}) \pmod 2$$

$$\equiv \sum_{B(k) \subseteq B(n)} x^k \pmod 2,$$

and it follows that $\binom{n}{k}$ is odd if and only if $B(k) \subseteq B(n)$.

Exercises for Section 1.3

1. There are several nice proofs of Fermat's Theorem. Fill in the details of the following proof that if p is a prime, then $a^p \equiv a \pmod p$ for every natural number a. This statement is clearly true for $a = 1$. The inductive step makes use of the following fact about binomial coefficients: if p is a prime, then

$$\binom{p}{k} \equiv 0 \pmod p, \quad k = 1, 2, \ldots, p-1.$$

Thus if $a^p \equiv a \pmod p$, then

$$(a+1)^p = a^p + \binom{p}{1} a^{p-1} + \cdots + \binom{p}{p-1} a + 1$$

$$\equiv a^p + 1$$

$$\equiv a + 1 \pmod p.$$

2. Prove that $20^{15} - 1$ is divisible by $11 \cdot 31 \cdot 61$.

3. Prove that
   (a) $19^{19} + 69^{69}$ is divisible by 44,
   (b) $2^{70} + 3^{70}$ is divisible by 13.

4. Find the last three digits of 13^{398}.

5. By Fermat's Theorem, we know that $n^5 \equiv n \pmod 5$. (a) Prove that $n^5 \equiv n \pmod{30}$. (b) Prove that if n is odd, then $n^5 \equiv n \pmod{240}$.

6. Prove that $(p-1)! \equiv -1 \pmod p$ for every prime p [Wilson's Theorem]. *Hint:* The statement is clearly true for $p = 2$ and $p = 3$. Assume that $p > 3$ and consider the set $M_p = \{1, 2, \ldots, p-1\}$. Prove that for every $a \in M_p$ there is a unique corresponding element $a' \in M_p$ such that $aa' \equiv 1 \pmod p$. Also prove that $a' = a$ only for $a = 1$ and $a = p - 1$. Thus excluding $a = 1$

and $a = p-1$, the remaining $p-3$ integers can be grouped into $(p-3)/2$ pairs such that the product of each pair is congruent to 1 (mod p). It follows that $(p-1)! \equiv -1 \pmod p$.

7. Prove that there is no integer n such that $n^2 + 3n + 4$ is divisible by 49. *Hint:* If there were such an integer, then it would satisfy $(2n+3)^2 + 7 \equiv 0 \pmod{49}$ and so $(2n+3)^2 \equiv 0 \pmod 7$.
8. Prove that $7|(a^2 + b^2)$ only when $7|a$ and $7|b$.
9. What powers of 2 give a remainder of 15 when divided by 17?
10. Determine $(n^3 + 1, n^2 + 2)$ for all n.
11. (a) Find all n for which $2^n - 1$ is divisible by 7.

    (b) Prove that there is no natural number n for which $2^n + 1$ is divisible by 7. [1964 IMO]
12. Set $a(1) = 1$ and $a(n) = n^{a(n-1)}$ for $n \geq 2$. Find the last three digits of $a(9)$. *Hint:* Use the heuristic of working backwards. What information concerning $a(8)$ is sufficient in order to determine the last three digits of $a(9)$? What information about $a(7)$ would tell you what you need to know about $a(8)$? Find the point at which you have the needed information and then work from that point forward to find the last three digits of $a(9)$.
13. What is the smallest natural number that leaves remainders $1, 2, 3, \ldots,$
    9 when divided by $2, 3, 4, \ldots, 10$, respectively? Find all such natural numbers.
14. Solve the congruence $x^3 \equiv 53 \pmod{120}$. *Hint:* Equivalently, find x such that

$$x^3 \equiv 2 \pmod 3,$$
$$x^3 \equiv 3 \pmod 5,$$
$$x^3 \equiv 5 \pmod 8.$$

Solve each of the above congruences, and then use the Chinese Remainder Theorem to put things together.

15. Let n be an arbitrary natural number. Prove that there is a pair of natural numbers a, b such that $(a + r, b + s) > 1$ for all $r, s = 1, 2, \ldots, n$. *Hint:* Introduce an $n \times n$ array of distinct primes

$$P = [p_{rs}], \qquad r, s = 1, 2, \ldots, n,$$

and show that there exists a pair a, b such that p_{rs} divides both $a+r$ and $b+s$ for all $r, s = 1, 2, \ldots, n$. Use the Chinese Remainder Theorem.

1.4 Rational and Irrational Numbers

The discovery of irrational numbers came as a big shock. The discovery was made by one of the Pythagoreans, and it is believed these followers of Pythagoras were so disturbed by the existence of such numbers that they threatened death to any member of the brotherhood who revealed the secret. Nowadays we are more casual about irrational numbers. Nevertheless, the distinction between rational and irrational numbers is the starting point for some rather deep mathematics, and many challenging and interesting problems have come from the Pythagorean discovery.

A **rational number** is one that can be expressed in the form a/b, where a and b are integers and $b \neq 0$. To represent a given nonzero rational number, we can choose a/b such that a is an integer, b is a natural number, and $(a, b) = 1$. We shall then say that the representative fraction is **in lowest terms**. An easy consequence of the definition is that any rational number has a periodic decimal expansion. This follows from the fact that when we divide by b the possible remainders are $0, 1, \ldots, b-1$. Eventually the whole process repeats. Real numbers with nonrepeating decimal expansions cannot be expressed in the form a/b and such numbers are called **irrational**.

To put the definition another way, a rational number is one that satisfies an equation of the form $c_0x + c_1 = 0$ where c_0 and c_1 are integers and $c_0 \neq 0$. More generally, a number that satisfies an equation of the form

$$c_0x^n + c_1x^{n-1} + \cdots + c_{n-1}x + c_n = 0,$$

where $c_0, \ldots, c_n$ are integers and $c_0 \neq 0$ is called **algebraic**. A number that is not algebraic is called **transcendental**. It is known that

$$\pi = 3.14159265358979323846\ldots$$

and

$$e = 2.71828182845904523536\ldots$$

are transcendental numbers.

A well-known and easily proved fact is that if $k \geq 2$ and n are natural numbers, then $\sqrt[k]{n}$ is irrational unless n is a perfect kth power. For assume that

$$\sqrt[k]{n} = \frac{a}{b}$$

where n is not a perfect kth power. Then $a^k = nb^k$, and some prime divisor of n has an exponent that is not a multiple of k. Let p be such a prime and note that the exponent of p in a^k is a multiple of k, but the exponent of p in $b^k n$ is not a multiple of k. This violates the Fundamental Theorem of Arithmetic, so our assumption that n is not a perfect kth power and $\sqrt[k]{n}$ is rational must be false.

One of the most useful facts to know for proving that certain numbers are irrational is the following basic theorem.

Theorem 1.10 (Rational Root Theorem) *If $c_0, c_1, \ldots, c_n$ are integers, a/b is in lowest terms and $x = a/b$ is a root of the equation*

$$c_0x^n + c_1x^{n-1} + \cdots + c_{n-1}x + c_n = 0,$$

then $a|c_n$ and $b|c_0$.

Indeed, if $x = a/b$ is a root of this equation, then

$$c_0a^n + c_1a^{n-1}b + \cdots + c_{n-1}ab^{n-1} + c_nb^n = 0.$$

Since a divides each of the other terms, it must divide c_nb^n. Since $(a, b) = 1$, Euclid's Lemma shows that $a|c_n$. In the same way we see that $b|c_0$.

Example 1.23 *Prove that $\sqrt{2} + \sqrt{3}$ is irrational.*

Solution. Set $x = \sqrt{2} + \sqrt{3}$. Squaring, we find $x^2 = 5 + 2\sqrt{6}$ and $(x^2 - 5)^2 = 24$. Thus x is a root of the equation $x^4 - 10x^2 + 1 = 0$. However, according to the Rational Root Theorem, the only possible rational roots of this equation are $x = \pm 1$. Neither 1 nor -1 satisfies the equation, so all its roots are irrational. In particular, $\sqrt{2} + \sqrt{3}$ is irrational, which is what we wanted to prove. □

Example 1.24 (1974 Putnam) *Prove that if* α *is a real number such that* $\cos(\alpha\pi) = 1/3$ *then* α *is irrational.*

Solution. In order to make a connection with the Rational Root Theorem, we first note some facts from trigonometry. From the addition formula for the cosine function, we have

$$\cos(n+1)\theta + \cos(n-1)\theta = 2\cos\theta\cos n\theta.$$

Thus if we set $x = 2\cos\theta$ and $P_n(x) = 2\cos n\theta$, then

$$P_{n+1}(x) = xP_n(x) - P_{n-1}(x),$$

with $P_0(x) = 2$ and $P_1(x) = x$. Thus $P_2(x) = x^2 - 2$, $P_3(x) = x^3 - 3x$, and so on. It follows by induction that for $n \geq 1$, $P_n(x)$ is a polynomial of degree n with integral coefficients and leading term x^n. (Polynomials whose highest power of x has coefficient 1 are called **monic**.) Suppose $\cos(m\pi/n) = a/b$ where a, b, m, n are integers. Then

$$P_n(2a/b) = 2\cos(m\pi) = 2(-1)^m,$$

so $x = 2a/b$ is a root of the equation

$$x^n + c_1x^{n-1} + \cdots + c_{n-1}x + c_n = 0,$$

where $c_1, \ldots, c_n$ are integers. By the Rational Root Theorem, $2a/b = 2\cos(m\pi/n)$ must be an integer, and since $|\cos\theta| \leq 1$ for all θ, the only possibilities are $2a/b = 0, \pm1, \pm2$. Thus $0, \pm1/2, \pm1$ are the only possible rational values for $\cos(\alpha\pi)$ if α is rational. In particular, if $\cos(\alpha\pi) = 1/3$ then α is irrational. □

There are many interesting questions having to do with **approximation** of irrationals by rationals.

Example 1.25 (1949 Putnam) *Let* a/b *(in lowest terms) represent a rational number that lies in the open interval* $(0, 1)$. *Prove that*

$$\left|\frac{a}{b} - \frac{\sqrt{2}}{2}\right| > \frac{1}{4b^2}.$$

Solution. Since $\sqrt{2}$ is irrational, we know that $2a^2 - b^2 \neq 0$. But $2a^2 - b^2$ is an *integer*, so in fact $|2a^2 - b^2| \geq 1$. It follows that

$$\left|\left(\frac{a}{b} - \frac{\sqrt{2}}{2}\right) \cdot \left(\frac{a}{b} + \frac{\sqrt{2}}{2}\right)\right| = \frac{|2a^2 - b^2|}{2b^2} \geq \frac{1}{2b^2}.$$

But a/b and $\sqrt{2}/2$ are each in $(0, 1)$, so

$$\frac{a}{b} + \frac{\sqrt{2}}{2} < 2.$$

Consequently,

$$\left|\frac{a}{b} - \frac{\sqrt{2}}{2}\right| > \frac{1}{4b^2}. \quad \square$$

Our final example is again concerned with rational roots of a polynomial equation. Its solution involves more than just an application of the Rational Root Theorem. In particular, the Factor Theorem and other basic facts concerning polynomial equations are used and the reader may find it advantageous to refer to §2.1 and §2.2, where these matters are dealt with in detail.

Example 1.26 (1981 British Olympiad) *Prove that if c is a rational number, the equation*

$$x^3 - 3cx^2 - 3x + c = 0$$

has at most one rational root.

Solution. It is worth noting that for certain values of c the equation has no rational root. As an example, if $c = 2$ the possible rational roots are limited to $\pm 1, \pm 2$ and it is easy to check that none of these satisfies the equation. On the other hand, for every rational number r there is some value of c so that r is a root of the equation. Specifically, if r is any rational number and we set

$$c = \frac{r(3 - r^2)}{1 - 3r^2},$$

then r is a root of the equation $x^3 - 3cx^2 - 3x + c = 0$. In this case, we must prove that there are no other rational roots. Let the remaining roots of the equation be s and t. Then

$$x^3 - 3cx^2 - 3x + c = (x - r)(x - s)(x - t),$$

and by comparing coefficients, we have

$$\begin{aligned} r + s + t &= 3c, \\ rs + rt + st &= -3 \\ rst &= -c. \end{aligned}$$

Since $s + t = 3c - r$ is rational, we know that either s and t are both rational or else they are both irrational. To prove that s and t are both irrational, note that from

$$3 = -\frac{r+s+t}{rst} = \frac{r+s-t}{r[3+r(s-t)]}$$

and $st = -3 - r(s+t)$ we get

$$s + t = \frac{8r}{1-3r^2} \quad \text{and} \quad st = \frac{r^2-3}{1-3r^2}.$$

It follows that

$$(s-t)^2 = (s+t)^2 - 4st = \frac{12(r^2+1)^2}{(1-3r^2)^2}.$$

Since this equation is equivalent to

$$\left\{\frac{(1-3r^2)(s-t)}{2(r^2+1)}\right\}^2 = 3$$

and $\sqrt{3}$ is irrational, it cannot be that s and t are rational. Thus we have proved that if c is a rational number, there is at most one rational root of the equation $x^3 - 3cx^2 - 3x + c = 0$. □

Exercises for Section 1.4

1. In the process of expanding the rational number a/b, where $0 < b < 100$, a student obtained the block of digits 143 somewhere beyond the decimal point. Prove that he made a mistake.
2. Prove that each of the following numbers is irrational.
   (a) $\sqrt{2} + \sqrt[3]{3}$
   (b) $\log_{10} 2$
   (c) $\pi + \sqrt{2}$
   Hint: To prove (c), show that if $x + \sqrt{2}$ is rational, then x is algebraic. Since it is known that π is transcendental, one concludes that $\pi + \sqrt{2}$ is irrational.

3. Show that $x = 2\cos(\pi/7)$ satisfies the equation

$$x^3 + x^2 - 2x - 1 = 0.$$

Using this fact, show that $\cos(\pi/7)$ is irrational.

4. Prove that there are no rational numbers a, b such that $1 + \sqrt{3} = (a + b\sqrt{3})^2$. What if $1 + \sqrt{3}$ is replaced by $37 + 20\sqrt{3}$?
5. Suppose that $x = 0.d_1d_2d_3\ldots$ where for $n > 2$ the digit d_n is the units digit of $d_{n-2} + d_{n-1}$. For example, if $d_{n-2} = 7$ and $d_{n-1} = 8$, then $d_n = 5$. Prove that x is rational.
6. Prove that if a, b, c and $\sqrt{a} + \sqrt{b} + \sqrt{c}$ are positive rational numbers, then $\sqrt{a}$, $\sqrt{b}$, and $\sqrt{c}$ are rational.
7. Prove that there is no set of integers m, n, p except 0, 0, 0 for which $m + n\sqrt{2} + p\sqrt{3} = 0$. [1955 Putnam]
8. Suppose that u, v, and uv are roots of the cubic equation

$$x^3 + ax^2 + bx + c = 0,$$

where a, b, and c are rational. Prove that if uv is irrational then $a = 1$.

9. Show that if r is a nonnegative rational approximation to $\sqrt{2}$, then $(r + 2)/(r + 1)$ is a *better* approximation.
10. Prove that if α and β are positive irrational numbers satisfying $1/\alpha + 1/\beta = 1$, then the sequences

$$\lfloor\alpha\rfloor,\ \lfloor 2\alpha\rfloor, \ldots,\ \lfloor k\alpha\rfloor, \ldots$$

and

$$\lfloor\beta\rfloor,\ \lfloor 2\beta\rfloor, \ldots,\ \lfloor k\beta\rfloor, \ldots$$

together include every natural number exactly once. (This result is known as Beatty's Theorem.) *Hint:* Let $\{x\} = x - \lfloor x\rfloor$ denote the fractional part of x. Show that if r is a real number greater than 1 and n is a natural number, then n is a term in the sequence $\lfloor r\rfloor$, $\lfloor 2r\rfloor, \ldots$ if and only if

$$\{n/r\} > 1 - \frac{1}{r}.$$

Show that if α and β are positive irrational numbers satisfying $1/\alpha + 1/\beta = 1$ then for every natural number n,

$$\{n/\alpha\} + \{n/\beta\} = 1,$$

and exactly one of the two inequalities

$$\{n/\alpha\} > 1 - \frac{1}{\alpha}, \quad \{n/\beta\} > 1 - \frac{1}{\beta}$$

holds.

1.5 Complex Numbers

The proclamation

Mathematicians: We're Number $-e^{i\pi}$

may not apply to *all* mathematicians, but it certainly applies to the person who gave us the symbol i for $\sqrt{-1}$, introduced e to represent 2.7182818..., standardized the symbol π, and discovered the amazing formula

$$e^{i\theta} = \cos\theta + i\sin\theta,$$

of which

$$-e^{i\pi} = 1$$

is a special case. Leonhard Euler was the most prolific mathematician of all time. It was said of Euler that he "calculated without apparent effort, as men breathe, or eagles sustain themselves in the air." Along with Johann Bernoulli, Abraham De Moivre and other leading mathematicians of the time, Euler made heroic calculations with complex numbers, demonstrating the undeniable usefulness of the numbers Descartes had called "imaginary."

Complex numbers extend the reals in a natural and useful way. The numbers in question are of the form $x + iy$ where x and y are real numbers, and their addition and multiplication are defined by

$$(x_1 + iy_1) + (x_2 + iy_2) = (x_1 + x_2) + i(y_1 + y_2)$$

and

$$(x_1 + iy_1)(x_2 + iy_2) = (x_1x_2 - y_1y_2) + i(x_1y_2 + x_2y_1),$$

respectively. Every complex number other than 0 has a multiplicative inverse

$$(x + iy)^{-1} = \frac{x}{x^2 + y^2} - i\frac{y}{x^2 + y^2}.$$

We shall use $\mathbb{C}$ to denote the set of all complex numbers. With the operations of addition and multiplication as defined, $\mathbb{C}$ is a field. If $z = x + iy$, we say that x is the **real part** of z and write

$$x = \operatorname{Re} z.$$

Similarly, y is the **imaginary part** of z, written

$$y = \operatorname{Im} z.$$

The **complex conjugate** of $z = x + iy$ is the complex number $\bar{z} = x - iy$. The **modulus** of z is the nonnegative real number

$$|z| = \sqrt{z\bar{z}} = \sqrt{x^2 + y^2}.$$

The complex number $z = x + iy$ can be associated with the point (x, y) in the Cartesian plane. Then $|z|$ is the **distance** from z to the origin, and more generally, $|z - w|$ is the distance from z to w. By introducing polar coordinates (r, θ), we can write

$$x + iy = r(\cos\theta + i\sin\theta) = re^{i\theta},$$

where $r = \sqrt{x^2 + y^2}$ and θ is chosen so that $x = r\cos\theta$ and $y = r\sin\theta$. (This determines θ up to a multiple of 2π.) Then r is the modulus of z, and for z nonzero, θ is an **argument** of z. The polar form is particularly convenient for multiplication. If $z_1 = r_1(\cos\theta_1 + i\sin\theta_1)$ and $z_2 = r_2(\cos\theta_2 + i\sin\theta_2)$, then

$$z_1z_2 = r_1r_2[\cos(\theta_1 + \theta_2) + i\sin(\theta_1 + \theta_2)].$$

(Multiply the moduli and add the arguments.) Specializing the above formula to the case $r_1 = r_2 = 1$ and using mathematical induction on n, one can easily prove the following important result.

Theorem 1.11 (De Moivre) *For every natural number n,*

$$(\cos\theta + i\sin\theta)^n = \cos n\theta + i\sin n\theta.$$

The formula of De Moivre immediately extends to all integers n since

$$\begin{aligned}(\cos\theta + i\sin\theta)^{-1} &= \cos\theta - i\sin\theta \\ &= \cos(-\theta) + i\sin(-\theta).\end{aligned}$$

Historical Note. The formulas of Euler and De Moivre were known by 1750, and at that time Euler had visualized complex numbers as points in the Cartesian plane. However, it was not until the early nineteenth century with the work of Argand and Gauss that the idea of the complex plane was firmly established and the arithmetic of complex numbers was described geometrically.

Using De Moivre's Theorem, we can develop an understanding of **roots** of complex numbers, starting with **roots of unity**. To say that $z = r(\cos\theta + i\sin\theta)$ is an nth root of unity means that $z^n = 1$, and this requires

$$r^n(\cos n\theta + i\sin n\theta) = 1.$$

To satisfy this requirement, we need $r = 1$ and $\cos n\theta = 1$. The latter is satisfied if and only if $\theta = 2k\pi/n$, where k is an integer. It follows that every nth root of unity is of the form ϵ^k where

$$\epsilon = \cos(2\pi/n) + i\sin(2\pi/n) = e^{2\pi i/n}.$$

A complete set of distinct roots is then

$$\{\epsilon^r \mid r = 0, 1, \ldots, n-1\}.$$

These numbers are certainly distinct. To see that they constitute a complete set of roots, just note that for every integer k there exist integers q and r with $0 \le r \le n-1$ such that $k = nq + r$. Then $\epsilon^k = \epsilon^r$ where $0 \le r \le n-1$, so every nth root of unity is included in the above set. The set of nth roots of unity has a beautiful description as a set of points in the complex plane. The roots are the vertices of a regular polygon with n sides inscribed in the circle $|z| = 1$. As an example, the cube roots of unity are 1, ω and $\omega^2 = \overline{\omega}$, where

$$\omega = \cos(2\pi/3) - i\sin(2\pi/3) = -\frac{1}{2} + i\frac{\sqrt{3}}{2}.$$

In the complex plane, the three roots 1, ω, $\overline{\omega}$ are the vertices of an equilateral triangle.

Example 1.27 *Find all fifth roots of unity.*

Solution. To express the roots of $z^5 = 1$ in polar form, we just apply the preceeding discussion to obtain the complete set of roots:

$$\left\{ \cos\left(\frac{2\pi k}{5}\right) + i \sin\left(\frac{2\pi k}{5}\right) \middle| \, k = 0, 1, 2, 3, 4 \right\}.$$

On the other hand, let us find the roots directly by factoring the equation

$$z^5 - 1 = 0.$$

First of all, $z^5 - 1 = (z-1)(z^4 + z^3 + z^2 + z + 1)$. Secondly,

$$z^4 + z^3 + z^2 + z + 1 = \left(z^2 + \frac{1}{2}z + 1\right)^2 - \frac{5}{4}z^2.$$

Hence the complex fifth roots of unity satisfy

$$\left(z^2 + \frac{1+\sqrt{5}}{2}z + 1\right)\left(z^2 + \frac{1-\sqrt{5}}{2}z + 1\right) = 0.$$

Thus we find the complete set of roots of $z^5 - 1 = 0$ to be

$$\left\{1, \frac{-(1+\sqrt{5}) \pm i\sqrt{10 - 2\sqrt{5}}}{4}, \frac{-(1-\sqrt{5}) \pm i\sqrt{10 + 2\sqrt{5}}}{4}\right\}.$$

By comparing the two forms of the solution, we deduce

$$\cos(2\pi/5) = \frac{\sqrt{5} - 1}{4}$$

and related special values of sine and cosine. □

Note. The preceeding solution leads directly to a mathematical problem of great historical importance. For which n is it possible to express all of the nth roots of unity in terms of a finite number of rational operations and the extraction of real square roots? It turns out that this question has the same answer as one that was raised by the Greeks: for which n is it possible to construct a regular polygon with n sides using only ruler and compass? Using the equivalence of these two questions, the preceding example shows that a ruler and compass construction exists for the regular pentagon. The Greeks already knew this and a good bit more. For example, Euclid knew that a ruler and compass construction exists in the case $n = 15$. However, it was not until Gauss that the complete story was told. Just before his nineteenth birthday, Gauss made the first advance since the time of Euclid by proving that a regular 17-gon was constructible. But he did much more. He proved that a regular n-gon was constructible if and only if the odd prime factors of n are all of the form $2^m + 1$ and none is repeated. (It is easy to prove that $2^m - 1$ can be prime only if m is a power of two. The only primes of this form that are known are 3, 5, 17, 257, 65537.) The example set by Gauss in solving a 2000 year old problem while still a teenager serves as a model for gifted young mathematicians.

Having found the nth roots of unity using De Moivre's formula, it is only a small step to find the nth roots of an arbitrary complex number. Suppose that c is a nonzero complex number, and we want to find all solutions of the equation

$$z^n = c.$$

Write $z = r(\cos\theta + i\sin\theta)$ and $c = \rho(\cos\alpha + i\sin\alpha)$ and use De Moivre's formula. Thus $z^n = c$ is true if and only if

$$r^n = \rho, \qquad \cos(n\theta) = \cos(\alpha), \qquad \sin(n\theta) = \sin(\alpha).$$

It follows as before that the solution set of $z^n = c$ is

$$\left\{ \sqrt[n]{\rho}\left[\cos\left(\frac{\alpha + 2\pi k}{n}\right) + i\sin\left(\frac{\alpha + 2\pi k}{n}\right)\right] \,\middle|\, k = 0, \ldots, n-1 \right\}.$$

Of course, if $c = 0$ then $z = 0$ is the only solution. Thus we know that for every complex number c, there exists a complex number z that satisfies $z^n - c = 0$. This result naturally leads to the following question: is it true that in the complex number system *every* polynomial equation of degree at least one has a solution? Euler was convinced that the answer is "yes" but he couldn't prove it. In fact, false proofs were given by some fairly famous mathematicians. The first correct proof was given by Gauss. He gave four proofs, the first being in his doctoral dissertation.

Theorem 1.12 (Fundamental Theorem of Algebra) *If $n \geq 1$ and c_0, $c_1, \ldots, c_n$ are complex numbers with $c_0 \neq 0$, there is a complex number z that satisfies the equation*

$$c_0 z^n + c_1 z^{n-1} + \cdots + c_n = 0.$$

There is no strictly algebraic proof of the Fundamental Theorem of Algebra. Known proofs use the methods of analysis, in particular theorems involving the concept of continuity.

Some special polynomial equations can be solved easily using our knowledge concerning the nth roots of unity.

Example 1.28 *Find all solutions of the equation* $(z-1)^n = z^n$.

Solution. For $z \neq 0$, the given equation is equivalent to

$$\left(\frac{z-1}{z}\right)^n = 1.$$

It follows that if z satisfies the equation, then $(z-1)/z$ is an nth root of unity. Since $(z-1)/z \neq 1$, the possibilities are

$$\frac{z-1}{z} = \epsilon^{2k}, \quad k = 1, 2, \ldots, n-1,$$

where

$$\epsilon = \cos(\pi/n) + i\sin(\pi/n).$$

Solving for z, we find

$$\begin{aligned} z &= \frac{-1}{\epsilon^{2k}-1} \\ &= \frac{1}{2}\left\{1 - \frac{\epsilon^k + \epsilon^{-k}}{\epsilon^k - \epsilon^{-k}}\right\} \\ &= \frac{1}{2}\left\{1 + i\cot\left(\frac{k\pi}{n}\right)\right\}, \quad k = 1, 2, \ldots, n-1. \quad \square \end{aligned}$$

Some real fun with complex numbers starts with the realization that many familiar formulas hold for complex numbers just as well as for reals. In the next section, we apply this observation to progressions. Here let us reap the benefit from knowing that the **Binomial Theorem** holds for complex numbers.

Theorem 1.13 (Binomial Theorem) *If n is any natural number and a and b are arbitrary real or complex numbers, then*

$$(a+b)^n = \sum_{k=0}^{n} \binom{n}{k} a^{n-k} b^k.$$

Example 1.29 *Prove the trigonometric identity*

$$\cos^n\theta = \frac{1}{2^n}\sum_{k=0}^{n}\binom{n}{k}\cos\{(n-2k)\theta\}.$$

Solution. Before showing how this identity follows from Euler's formula and the Binomial Theorem, let us pause for a moment to look at some special cases. Writing out the identity for $n = 2, 3$, and 4 and using the fact that $\cos(-x) = \cos(x)$ to simplify the resulting expression, we find

$$\begin{aligned}\cos^2\theta &= (1+\cos 2\theta)/2,\\ \cos^3\theta &= (3\cos\theta + \cos 3\theta)/4,\\ \cos^4\theta &= (3 + 4\cos 2\theta + \cos 4\theta)/8.\end{aligned}$$

These formulas have many useful applications. Now to the proof:

$$\begin{aligned}\cos^n\theta &= \left(\frac{e^{i\theta}+e^{-i\theta}}{2}\right)^n\\ &= \frac{1}{2^n}\sum_{k=0}^{n}\binom{n}{k}(e^{i\theta})^{n-k}(e^{-i\theta})^k\\ &= \frac{1}{2^n}\sum_{k=0}^{n}\binom{n}{k}e^{i(n-2k)\theta}\\ &= \frac{1}{2^n}\sum_{k=0}^{n}\binom{n}{k}\cos\{(n-2k)\theta\}.\end{aligned}$$

In the last step, we have again used Euler's formula along with the realization that only the real part of each term contributes since the sum is real. (The fact that the imaginary parts all cancel out is obvious since $\binom{n}{k} = \binom{n}{n-k}$ and $\sin(-x) = -\sin(x)$.) □

Two immediate consequences of the Binomial Theorem are

$$\sum_{k=0}^{n}\binom{n}{k} = (1+1)^n = 2^n, \qquad \sum_{k=0}^{n}\binom{n}{k}(-1)^k = (1-1)^n = 0 \quad (n > 0).$$

Adding these two equations and dividing by 2 we obtain

$$\sum_{k\equiv 0 \pmod 2}\binom{n}{k} = 2^{n-1}.$$

and thus

$$\sum_{k \equiv 1 \pmod 2} \binom{n}{k} = 2^{n-1}$$

for $n > 0$. This example begs to be generalized. Thus we seek a general formula by which to compute

$$\sum_{k \equiv r \pmod m} \binom{n}{k}.$$

This provides a job for complex numbers. We begin with a special case.

Example 1.30 *Show that*

$$\sum_{k \equiv 0 \pmod 3} \binom{n}{k} = \frac{1}{3}\left[2^n + 2\cos\left(\frac{n\pi}{3}\right)\right].$$

Solution. Let $\omega = \frac{-1+i\sqrt{3}}{2}$. Then we claim that

$$1 + \omega^k + \omega^{2k} = \begin{cases} 3 & \text{if } k \equiv 0 \pmod 3, \\ 0 & \text{otherwise.} \end{cases}$$

To see this, simply note that since $\omega^3 = 1$,

$$(1 - \omega^k)(1 + \omega^k + \omega^{2k}) = 1 - \omega^{3k} = 0.$$

If $k \equiv 0 \pmod 3$, then $1 + \omega^k + \omega^{2k} = 3$. Otherwise, $\omega^k \neq 1$ and the above equation shows that $1 + \omega^k + \omega^{2k} = 0$. Using $\omega^2 = \overline{\omega}$ and $1 + \omega = \cos(\pi/3) + i\sin(\pi/3)$, we find

$$\begin{aligned} \sum_{k \equiv 0 \pmod 3} \binom{n}{k} &= \frac{1}{3}\sum_{k=0}^{n} \binom{n}{k}(1 + \omega^k + \overline{\omega}^k) \\ &= \frac{1}{3}\left[2^n + (1+\omega)^n + (1+\overline{\omega})^n\right] \\ &= \frac{1}{3}\left[2^n + 2\cos\left(\frac{n\pi}{3}\right)\right]. \quad \square \end{aligned}$$

With the idea of this example in mind, it is not difficult to find the extension that allows us to sum the terms belonging to a given residue class. We can build a "residue class selector" (mod m) using the complex mth root of unity $\epsilon = e^{2\pi i/m}$. The desired formula is

$$\frac{1}{m}\sum_{s=0}^{m-1} \epsilon^{s(k-r)} = \begin{cases} 1 & \text{if } k \equiv r \pmod{m}, \\ 0 & \text{otherwise.} \end{cases}$$

From this we obtain the following useful result.

Theorem 1.14 (Multisection Formula) *If* $f(x) = \sum_k a_k x^k$, *then*

$$\sum_{k \equiv r \pmod{m}} a_k x^k = \frac{1}{m}\sum_{s=0}^{m-1} \epsilon^{-rs} f(\epsilon^s x),$$

where $\epsilon = e^{2\pi i/m}$.

Note. We have been intentionally vague concerning the summation range in the definition of f. In fact, this formula holds whenever the sum defining $f(x)$ converges. Thus it certainly holds if f is a polynomial, but it may also hold when the sum defining f is an infinite series.

Using the multisection formula together with that of De Moivre, we can offer the following solution of the problem posed earlier:

$$\sum_{k \equiv r \pmod{m}} \binom{n}{k} = \frac{1}{m}\sum_{s=0}^{m-1} \cos\left((n-2r)\frac{s\pi}{m}\right) 2^n \cos^n\left(\frac{s\pi}{m}\right).$$

Our final example uses a veritable "kitchen sink" of mathematical ideas (congruence, the binomial expansion, complex numbers) to solve a divisibility problem.

Example 1.31 (1974 IMO) *Prove that the number*

$$\sum_{k=0}^{n} \binom{2n+1}{2k+1} 2^{3k}$$

is not divisible by 5 for any integer $n \geq 0$.

Solution. Since $2^3 \equiv -2 \pmod{5}$, an equivalent problem is to prove that

$$S_n = \sum_{k=0}^{n} \binom{2n+1}{2k+1} (-2)^k$$

is not divisible by 5. We expand $(1+i\sqrt{2})^{2n+1}$ using the Binomial Theorem and separate the even and odd terms to get

$$(1+i\sqrt{2})^{2n+1} = R_n + i\sqrt{2}\,S_n, \tag{1.5}$$

where

$$R_n = \sum_{k=0}^{n} \binom{2n+1}{2k} (-2)^k.$$

Multiplying each side of (1.5) by its conjugate, we obtain

$$3^{2n+1} = R_n^2 + 2S_n^2.$$

Since $3^2 \equiv -1 \pmod 5$, the last equation leads to

$$\pm 3 \equiv R_n^2 + 2S_n^2 \pmod 5,$$

and this tells the story. For $S_n \equiv 0 \pmod 5$ implies that $R_n^2 \equiv \pm 3 \pmod 5$, a contradiction since any square is congruent to 0, 1, or 4 $\pmod 5$. Thus $S_n \equiv 0 \pmod 5$ is impossible. □

Exercises for Section 1.5

1. If z and w are nonzero complex numbers, the points $0,\ z,\ w,\ z+w$ are the vertices of a parallelogram in the complex plane. Show that

$$|z+w|^2 + |z-w|^2 = 2\{|z|^2 + |w|^2\}.$$

Give a geometrical interpretation of this result.

2. Find the three cube roots of $-i$.

3. Solve each of the following equations:

(a) $|z| + z = 2 + i$,

(b) $z^2 - z + i\bar{z} = 0$.

4. Find all z that satisfy $\operatorname{Re} z = \operatorname{Im} z$ and

$$|z| + |z+14| = 28.$$

5. Show that the roots of the equation

$$z^3 + c_1 z^2 + c_2 z + c_3 = 0$$

are the vertices of an equilateral triangle (possibly degenerating to a point) if and only if $3c_2 = c_1^2$.

6. Show that if a, b, c are complex numbers satisfying $|a| = |b| = |c| = r$ where $r > 0$, then

$$\left|\frac{ab + bc + ca}{a + b + c}\right| = r.$$

7. Let z be any complex number with modulus 1 other than $z = -1$. Show that there is a real number t such that

$$z = \frac{1 + it}{1 - it}.$$

8. Evaluate

$$\sum_{k \equiv 1 \ (\mathrm{mod}\ 3)} \binom{n}{k} \quad \text{and} \quad \sum_{k \equiv 2 \ (\mathrm{mod}\ 3)} \binom{n}{k}.$$

9. Show that

$$\sum_{k=0}^{n} \binom{n}{k} \cos(k\theta) = \left(2\cos\frac{\theta}{2}\right)^n \cos\left(\frac{n\theta}{2}\right).$$

10. Find all solutions of the equation $(z^2 - 1)^3 = 1$.

11. Show that all of the roots of the equation

$$(z - 1)^5 = 32(z + 1)^5$$

lie on the circle of radius $4/3$ centered at $(-5/3, 0)$.

12. Find all solutions of the equation $(1 + z)^{2n} + (1 - z)^{2n} = 0$.

13. Verify that the solution set of

$$z^5 + z^4 - 4z^3 - 3z^2 + 3z + 1 = 0$$

is

$$\left\{ 2\cos\left(\frac{2k\pi}{11}\right) \middle| \, k = 1, 2, 3, 4, 5 \right\}.$$

Hint: Set $z = w + w^{-1}$ and show that z satisfies the above equation if and only if $w \neq 1$ is a complex 11th root of unity.

14. Fix one vertex of a regular n-gon inscribed in the unit circle and consider the $n - 1$ diagonals that join this vertex to each of the remaining $n - 1$ vertices of the n-gon. Prove that the product of the lengths of these diagonals is n. *Hint:* We can take the vertices

of the regular n-gon to be the nth roots of unity and the fixed vertex to be $(1, 0)$. Thus we want to compute

$$|1-\epsilon| \cdot |1-\epsilon^2| \cdots |1-\epsilon^{n-1}|,$$

where

$$\epsilon = \cos(2\pi/n) + i\sin(2\pi/n).$$

Show that this is simply $|P(1)|$, where $P(z) = z^{n-1} + z^{n-2} + \cdots + z + 1$.

15. Show that

$$\sin\left(\frac{2\pi}{7}\right) + \sin\left(\frac{4\pi}{7}\right) + \sin\left(\frac{8\pi}{7}\right) = \frac{\sqrt{7}}{2}.$$

Hint: Let $\alpha = \cos\left(\frac{2\pi}{7}\right) + i\sin\left(\frac{2\pi}{7}\right)$, $p = \alpha + \alpha^2 + \alpha^4$, and $q = \alpha^3 + \alpha^5 + \alpha^6$. Then the desired quantity is $\operatorname{Im} p$. Show that $p + q = -1$ and $pq = 2$.

1.6 Progressions and Sums

It was known in antiquity that if $a_1, a_2, \ldots, a_n$ are in arithmetic progression, then

$$a_1 + a_2 + \cdots + a_n = n\left(\frac{a_1 + a_n}{2}\right).$$

However, it is one thing for a formula to be known by practicing mathematicians and quite another for it to be deduced in an instant by a ten-year-old boy. This is exactly what Gauss did when his arithmetic teacher, Herr Büttner, gave Gauss and his classmates a problem specifically designed to keep them hard at work for an hour. The problem chosen to create tedium and frustration was that of summing an arithmetic progression. Immediately Gauss wrote a number on his slate, turned it in and announced, "There it is." At the end of the hour, the number written by Gauss was the only correct answer to come from the class. What Gauss immediately recognized was that in an arithmetic progression $a_1, a_2, \ldots, a_n$,

$$a_1 + a_n = a_2 + a_{n-1} = a_3 + a_{n-2} = \cdots,$$

so the sum is the same as if every one of the n terms had the "average" value $(a_1 + a_n)/2$.

An **arithmetic progression** is a sequence $a_1, a_2, a_3 \ldots$ (finite or infinite) with the property that $a_2 - a_1 = a_3 - a_2 = \cdots = d$; the parameter d is then called the **common difference**. Alternatively, the characteristic property is that each interior term of the sequence is the average of its predecessor and successor:

$$a_k = \frac{a_{k-1} + a_{k+1}}{2}.$$

An infinite arithmetic progression is characterized by its first term and common difference. (In the case of a finite arithmetic progression, characterization requires the number of terms as well.) For an arithmetic progression with initial term a_1 and common difference d, the kth term is

$$a_k = a_1 + (k-1)d.$$

Finding the sum of a finite arithmetic progression is child's play once the relevant parameters are identified.

Example 1.32 *Let the sequence of natural numbers be partitioned into groups as follows:*

$$1,\ (2,3),\ (4,5,6),\ (7,8,9,10),\ (11,12,13,14,15),\ldots$$

Find the sum of the integers in the nth group.

Solution. The number of terms in the first k groups is $1 + 2 + \cdots + k = k(k+1)/2$. Therefore the first of the n terms in the nth group is $(n-1)n/2 + 1$ and the last is $n(n+1)/2$. The average value of a term in the nth group is therefore

$$\frac{1}{2}\left(\frac{n^2 - n}{2} + 1 + \frac{n^2 + n}{2}\right) = \frac{n^2 + 1}{2},$$

and the desired sum is

$$S_n = n\left(\frac{n^2 + 1}{2}\right). \quad \square$$

The next example provides an opportunity to use what we know about arithmetic progressions in the context of a divisibility problem.

Example 1.33 *Prove that the powers of two are the only positive integers that cannot be written as the sum of two or more consecutive positive integers.*

Solution. Suppose $n = m + (m+1) + \cdots + (m+k)$ for some positive integers m and k. Then $n = (k+1)(2m+k)/2$. If k is odd then $(2m+k)|n$, and if k is even then $(k+1)|n$; in either case, n has an odd divisor greater than 1. Thus n is not a power of two. Conversely, suppose n is not a power of two. Then $n = 2^a(2b+1)$ where $b \geq 1$. If $b < 2^a$, then n is the sum of $2b+1$ positive terms with average value 2^a:

$$n = (2^a - b) + (2^a - b + 1) + \cdots + (2^a + b).$$

If $b \geq 2^a$, then n is the sum of 2^{a+1} positive terms with average value $(2b+1)/2$:

$$n = (b - 2^a + 1) + (b - 2^a + 2) + \cdots + (b + 2^a).$$

In either case, n can be written as the sum of two or more consecutive positive integers. □

Example 1.34 (1979 Putnam) *Let $x_1, x_2, x_3, \ldots$ be a sequence of nonzero real numbers satisfying*

$$x_n = \frac{x_{n-2}x_{n-1}}{2x_{n-2} - x_{n-1}}$$

for $n = 3, 4, 5, \ldots$ Establish a necessary and sufficient condition on x_1 and x_2 for x_n to be an integer for infinitely many values of n.

Solution. The recurrence formula may be written

$$\frac{1}{x_n} = \frac{2}{x_{n-1}} - \frac{1}{x_{n-2}}.$$

Substituting $r_n = 1/x_n$, we find that

$$r_{n-1} = \frac{r_{n-2} + r_n}{2},$$

and it follows that the sequence $r_1, r_2, \ldots$ is an arithmetic progression. The initial term of this progression is $r_1 = 1/x_1$ and the common difference is

$$d = \frac{1}{x_2} - \frac{1}{x_1} = \frac{x_1 - x_2}{x_1 x_2}.$$

Using $r_n = r_1 + (n-1)d$, we obtain

$$x_n = \frac{1}{r_n} = \frac{x_1 x_2}{x_2 + (n-1)(x_1 - x_2)}.$$

If $x_1 \neq x_2$, then $|x_n| < 1$ for all sufficiently large values of n, so in this case the sequence takes integral values only a finite number of times. The only way for x_n to be an integer for infinitely many n is for x_1 and x_2 to be equal. In this case, the sequence is constant ($x_n = x_1$ for all n) so we must require that x_1 be an integer. □

The sequence $a = a_1, a_2, a_3, \ldots$ is a **geometric progression** if there is a number r such that for $k = 1, 2, 3, \ldots$,

$$a_{k+1} = r\, a_k.$$

The parameter r is called the **common ratio** and the nth term in the sequence is given by

$$a_n = a\, r^{n-1}.$$

The sum of a finite geometric progression is easily obtained. If

$$S_n = a + ar + ar^2 + \cdots + ar^{n-1},$$

then

$$rS_n = ar + ar^2 + ar^3 + \cdots + ar^n,$$

so

$$(1-r)S_n = a - ar^n.$$

If $r \neq 1$ we can divide by $r - 1$ and solve for S_n. But if $r = 1$ the problem is trivial since every term is a and the sum is na. Thus we have

$$S_n = \begin{cases} \dfrac{a(1-r^n)}{1-r} & \text{if } r \neq 1, \\ an & \text{if } r = 1. \end{cases}$$

Example 1.35 *Suppose $a_1 = 2$ and $a_{k+1} = 3a_k + 1$ for all $k \geq 1$. Find a general formula for $a_1 + a_2 + \cdots + a_n$.*

Solution. Although the sequence (a_k) is not a geometric progression, it is simply related to one. Write the formula connecting a_k and a_{k+1}

as

$$a_{k+1} + \frac{1}{2} = 3\left(a_k + \frac{1}{2}\right).$$

Then the sequence (b_n) defined by $b_k = a_k + \frac{1}{2}$ is a geometric progression and we can apply the above formula to find the sum of the first n terms. We find $b_k = (5/2)3^{k-1}$ for the general term and

$$b_1 + b_2 + \cdots + b_n = \frac{5}{4}(3^n - 1)$$

for the sum. It follows that

$$a_1 + a_2 + \cdots + a_n = \frac{5}{4}(3^n - 1) - \frac{n}{2}. \quad \square$$

Example 1.36 *Show that*

$$\sin\left(\frac{\pi}{n}\right) + \sin\left(\frac{2\pi}{n}\right) + \cdots + \sin\left(\frac{(n-1)\pi}{n}\right) = \cot\left(\frac{\pi}{2n}\right).$$

Solution. We use De Moivre's Theorem and the fact that the imaginary part of a sum is the sum of the imaginary parts. Let $z = \cos(\pi/2n) + i\sin(\pi/2n)$. Then $\sin(k\pi/n) = \operatorname{Im} z^{2k}$ and $z^{2n} = -1$. By summing the appropriate geometric progression, we obtain

$$\begin{aligned}
\sum_{k=1}^{n-1} \sin\left(\frac{k\pi}{n}\right) &= \operatorname{Im} \sum_{k=1}^{n-1} z^{2k} \\
&= \operatorname{Im} \frac{z^2 - z^{2n}}{1 - z^2} \\
&= \operatorname{Im} \frac{z^2 + 1}{1 - z^2} \\
&= \operatorname{Im} \frac{z + z^{-1}}{z^{-1} - z} \\
&= \operatorname{Im} \frac{2\cos(\pi/2n)}{-2i\sin(\pi/2n)} \\
&= \cot(\pi/2n). \quad \square
\end{aligned}$$

The summation formulas for arithmetic and geometric progressions are solutions in special cases of the following general problem:

given a sequence $a_1, a_2, \ldots$, find a general formula for

$$S_n = \sum_{k=1}^{n} a_k.$$

Examples of this problem were solved in §1.2. On many occasions, it is helpful to have a formula by which to compute

$$S_r(n) = \sum_{k=1}^{n} k^r.$$

The following special cases are worth remembering:

$$\begin{aligned} S_1(n) &= n(n+1)/2, \\ S_2(n) &= n(n+1)(2n+1)/6, \\ S_3(n) &= (n(n+1)/2)^2. \end{aligned}$$

Note. The general result is

$$S_r(n) = \frac{B_{r+1}(n+1) - B_{r+1}(0)}{r+1},$$

where $B_m(x)$ is the **Bernoulli polynomial** of degree m. These polynomials were introduced by Jakob (James) Bernoulli. The mth Bernoulli polynomial satisfies the relation $B_m(x+1) - B_m(x) = mx^{m-1}$, and this fact yields the above summation formula.

Example 1.37 *Evaluate*

$$\sum_{k=1}^{n} \frac{1}{(k+1)\sqrt{k} + k\sqrt{k+1}}.$$

Solution. Note that

$$\begin{aligned} \frac{1}{(k+1)\sqrt{k} + k\sqrt{k+1}} &= \frac{1}{\sqrt{k(k+1)}(\sqrt{k+1} + \sqrt{k})} \\ &= \frac{\sqrt{k+1} - \sqrt{k}}{\sqrt{k(k+1)}} \\ &= \frac{1}{\sqrt{k}} - \frac{1}{\sqrt{k+1}}. \end{aligned}$$

Written out, the sum is

$$\left(1 - \frac{1}{\sqrt{2}}\right) + \left(\frac{1}{\sqrt{2}} - \frac{1}{\sqrt{3}}\right) + \cdots + \left(\frac{1}{\sqrt{n}} - \frac{1}{\sqrt{n+1}}\right),$$

and this collapses to give the result:

$$\sum_{k=1}^{n} \frac{1}{(k+1)\sqrt{k} + k\sqrt{k+1}} = 1 - \frac{1}{\sqrt{n+1}}. \quad \square$$

Sums like this (where almost everything cancels out) are called **telescoping**. Looking back at §1.2, you will see that the sum formulas we proved using the Principle of Mathematical Induction were precisely those for which we assumed a telescoping form $f(k) = F(k) - F(k-1)$.

Certain sums which are closely related to those of geometric progressions can be evaluated by similar methods.

Example 1.38 *Find a formula for the sum* $1 + 2r + 3r^2 + \cdots + nr^{n-1}$.

Solution. For $r = 1$ the sum is

$$1 + 2 + \cdots + n = n(n+1)/2.$$

Now assume that $r \neq 1$. If

$$S_n = 1 + 2r + \cdots + nr^{n-1},$$

then

$$rS_n = r + 2r^2 + \cdots + (n-1)r^{n-1} + nr^n,$$

and

$$\begin{aligned}(1-r)S_n &= 1 + (r + r^2 + \cdots + r^{n-1}) - nr^n \\ &= 1 - nr^n + \frac{r - r^n}{1-r}.\end{aligned}$$

Thus for $r \neq 1$,

$$S_n = \frac{1 - nr^n}{1-r} + \frac{r - r^n}{(1-r)^2}. \quad \square$$

Our final example involving arithmetic progressions provides a change of pace. In this problem we want to prove the absence of an arithmetic progression.

Example 1.39 (1972 Putnam) *Show that there are no four consecutive binomial coefficients*

$$\binom{n}{k}, \binom{n}{k+1}, \binom{n}{k+2}, \binom{n}{k+3}$$

(n, k positive integers and $\le k + 3 \le n$) which are in arithmetic progression.

Solution. The binomial coefficients fill a triangular array known as **Pascal's Triangle**, in which their computation is easily accomplished using the relation

$$\binom{r+1}{s+1} = \binom{r}{s} + \binom{r}{s+1}.$$

Suppose

$$\binom{n}{k}, \binom{n}{k+1}, \binom{n}{k+2}, \binom{n}{k+3}$$

are in arithmetic progression and consider the following portion of Pascal's triangle:

$$\begin{array}{cccc} \binom{n}{k} & \binom{n}{k+1} & \binom{n}{k+2} & \binom{n}{k+3} \\ & \binom{n+1}{k+1} & \binom{n+1}{k+2} & \binom{n+1}{n+3} \\ & & \binom{n+2}{k+2} & \binom{n+2}{k+3} \end{array}$$

Following our assumption, the entries can be filled in to produce the following array:

$$\begin{array}{cccc} a & a+d & a+2d & a+3d \\ & 2a+d & 2a+3d & 2a+5d \\ & & 4a+4d & 4a+8d \end{array}$$

Thus

$$\frac{\binom{n+2}{k+2}}{\binom{n}{k+1}} = \frac{\binom{n+2}{k+3}}{\binom{n}{k+2}} = 4.$$

Suppose the equality on the left holds. Then

$$\frac{1}{(k+2)(n-k)} = \frac{1}{(k+3)(n-k-1)},$$

which yields $n = 2k + 3$. But then the equality on the right doesn't hold:

$$\frac{\binom{2k+5}{k+2}}{\binom{2k+3}{k+1}} = \frac{2(2k+5)}{k+3} = 4 - \frac{2}{k+3} \neq 4.$$

Thus our assumption that

$$\binom{n}{k}, \binom{n}{k+1}, \binom{n}{k+2}, \binom{n}{k+3}$$

are in arithmetic progression has produced a contradiction. □

Exercises for Section 1.6

1. Find a general formula for the sum

$$1 + 11 + 111 + \cdots + 11\ldots1,$$

where the last term has n digits.

2. Find the sum of all the digits used in writing down the numbers from 1 to 1 billion (10^9).

3. Show that if a, b, c are positive numbers such that a^2, b^2, c^2 are in arithmetic progression, then

$$\frac{1}{b+c}, \frac{1}{c+a}, \frac{1}{a+b}$$

are in arithmetic progression.

4. For each n, find the sum of the integers between 1 and $10n$ that are not divisible by either 2 or 5.

5. Given that $a_1, a_2, \ldots, a_n$ are positive numbers in arithmetic progression, show that

$$\frac{1}{a_1a_2} + \frac{1}{a_2a_3} + \cdots + \frac{1}{a_{n-1}a_n} = \frac{n-1}{a_1a_n}$$

and

$$\frac{1}{\sqrt{a_1} + \sqrt{a_2}} + \cdots + \frac{1}{\sqrt{a_{n-1}} + \sqrt{a_n}} = \frac{n-1}{\sqrt{a_1} + \sqrt{a_n}}.$$

6. Let $a_1, a_2, a_3, \ldots$ be an arithmetic progression. Show that if $a_p = q$ and $a_q = p$ for some $p \neq q$, then $a_{p+q} = 0$.

7. Suppose that $a_1, a_2, \ldots, a_n$ are in arithmetic progression. Find a formula for $a_1^2 + a_2^2 + \cdots + a_n^2$ in terms of n, a_1, and a_n.
8. Let $k > 1$ and n be natural numbers. Show that there exist n consecutive odd integers whose sum is n^k.
9. Show that

$$\frac{1}{2} + \cos\theta + \cos 2\theta + \cdots + \cos n\theta = \frac{\sin\left(\frac{(2n+1)\theta}{2}\right)}{2\sin\left(\frac{\theta}{2}\right)}.$$

Hint: Let $z = e^{i\theta/2}$ and note that the sum in question can be written in terms of z as $\frac{1}{2}(z^{-2n} + z^{-(2n-2)} + \cdots + z^{2n})$. Summing the geometric progression, we obtain

$$\frac{1}{2}(z^{-2n} + z^{-(2n-2)} + \cdots + z^{2n}) = \frac{z^{-2n} - z^{2n+2}}{2(1 - z^2)}$$
$$= \frac{z^{2n+1} - z^{-(2n+1)}}{2(z - z^{-1})}.$$

10. Evaluate the sum

$$\frac{1}{1 \cdot 2} + \frac{2}{1 \cdot 3} + \cdots + \frac{F_{n+1}}{F_n \cdot F_{n+2}},$$

where F_k denotes the kth Fibonacci number (see Exercise 9, §1.2).
11. Evaluate the sum

$$\sum_{k=0}^{n}(F_{2^k})^{-1}.$$

Hint: By induction,

$$F_{k-1}F_m - F_kF_{m-1} = (-1)^kF_{m-k}, \qquad k = 1, 2, \ldots, m,$$

with the understanding that $F_0 = 0$. Setting $m = 2k$ we find

$$F_{k-1}F_{2k} - F_kF_{2k-1} = (-1)^kF_k,$$

so

$$\frac{(-1)^k}{F_{2k}} = \frac{F_{k-1}}{F_k} - \frac{F_{2k-1}}{F_{2k}}.$$

Thus the sum $\sum_{k=0}^{n}(F_{2^k})^{-1}$ telescopes.

12. Evaluate
$$\sum_{k=1}^{n} \frac{k}{k^4 + k^2 + 1}.$$

13. Evaluate
$$\sum_{k=0}^{r} (-1)^k \binom{n}{k}$$
for $0 \le r < n$.

14. Suppose that $a_1 = 1$ and $a_{k+1} = 2a_k + k$ for all $k \ge 1$. Find an explicit formula for a_n.

15. Find a general formula for
$$\sum_{k=0}^{n-1} (k+1) \cos\left(\frac{2\pi k}{n}\right).$$

1.7 Diophantine Equations

We do not know very much about the personal life of Diophantus of Alexandria, but some information about him is suggested in the following problem, which is found in the *Greek Anthology* (assembled around A.D. 500 by Metrodorus).

> "His boyhood lasted one sixth of his life; his beard grew after one twelfth more; he married after one seventh more; his son was born five years later; the son lived to half his father's age, and the father died four years after his son. How old was Diophantus when he died?"

One of the contributions of Diophantus was a famous book, the *Arithmetica*. In this book, Diophantus discusses the type of problem to which his name is now attached. The problem is to find all integral (or rational) solutions of an equation in two or more variables. Fourteen centuries later, Fermat was studying the *Arithmetica* when he produced the most famous problem involving such Diophantine equations, namely to show that for $n > 2$ the equation
$$x^n + y^n = z^n$$

has no solution where x, y, and z are natural numbers. Fermat wrote that he had found a truly remarkable proof, but that the margin (of the *Arithmetica*) was to small to contain it. Very recently, Andrew Wiles found a proof of Fermat's assertion.

Fortunately, not all Diophantine equations are as difficult as that of Fermat. Some are quite easy when looked at from an appropriate angle. However, there is no general and effective algorithm for dealing with Diophantine equations. Each one may be a new challenge to the problem solver's skill. It is desirable to have experience with several different strategies for dealing with Diophantine equations. Some of the simplest techniques are listed below.

1. Factoring
2. Congruence
3. Use of the Discriminant for Quadratic Equations
4. Fermat's Method of Infinite Descent
5. Special Forms

The following examples illustrate these methods. The first one uses a simple divisibility argument.

Example 1.40 *Find all pairs (x, y) of integers that satisfy the equation* $x(y + 1)^2 = 243y$.

Solution. Since $(y + 1, y) = 1$, it follows from Euclid's Lemma that $(y + 1)^2 | 243$. But since $243 = 3^5$, the only square divisors of 243 are 1, 9 and 81. Thus the possible values of y are $0, -2, 2, -4, 8, -10$. Computing $x = 243y/(y + 1)^2$ in each case, we find all pairs (x, y) that satisfy the given equation:

$$(0, 0), (-486, -2), (54, 2), (-108, -4), (24, 8), (-30, -10). \quad \square.$$

Often the fact that a Diophantine equation has no solution follows easily by a congruence argument.

Example 1.41 *Prove that the equation $x^2 = 3y^2 + 8$ has no solution in integers x, y.*

Solution. If there are integers x and y satisfying this equation then $x^2 \equiv 2 \pmod 3$, but this is impossible. (If $x \equiv 0 \pmod 3$ then $x^2 \equiv 0 \pmod 3$; otherwise, $x^2 \equiv 1 \pmod 3$.) $\square$

Here is another example where a congruence argument is key.

Example 1.42 (1976 USAMO) *Find all solutions of the equation* $a^2 + b^2 + c^2 = a^2b^2$ *in natural numbers* a, b, c.

Solution. Suppose that natural numbers a, b, c satisfy the equation. Write the equation as

$$c^2 = (a^2 - 1)(b^2 - 1) - 1,$$

and consider this equality modulo 4. If either a or b is odd, then the equation requires that $c^2 \equiv -1 \pmod 4$, which is impossible. Hence a and b are even, and so is c. Let $r \geq 1$ be the largest natural number such that 2^r divides each of a, b, c and write $a = 2^r x$, $b = 2^r y$, and $c = 2^r z$. Then at least one of the numbers x, y, z is odd and

$$x^2 + y^2 + z^2 = 2^{2r}x^2y^2.$$

Thus we come to $x^2 + y^2 + z^2 \equiv 0 \pmod 4$, where at least one of x, y, z is odd, and this is clearly impossible. Hence there are no solutions of the equation $a^2 + b^2 + c^2 = a^2b^2$ in natural numbers. □

Useful information concerning an equation that is quadratic in one of the variables can be obtained by using the fact that an integral solution is possible only if the discriminant is a perfect square.

Example 1.43 *Find all solutions of the equation* $x^2 - xy + y = 3$ *in integers* x, y.

Solution. This is a quadratic equation for x, so an integral solution implies that the discriminant, $y^2 - 4(y - 3) = (y - 2)^2 + 8$, is a perfect square, say z^2. Now $z^2 - (y - 2)^2 = (z - y + 2)(z + y - 2) = 8$ is a factorization of 8 into two factors whose sum is $2z$. It follows that the factorization is either $8 = 4 \cdot 2$ or $8 = (-4) \cdot (-2)$, so $z = \pm 3$. Thus $y - 2 = \pm 1$, so $y = 1$ and $y = 3$ are the only possible values. Substituting these values of y into the equation, we obtain $x^2 - x - 2 = 0$ and $x^2 - 3x = 0$, respectively. Each yields two solutions, and thus we obtain all solutions of the original equation: $(2, 1)$, $(-1, 1)$, $(0, 3)$, $(3, 3)$. □

The following problem involving divisibility is one of the hardest to have appeared on an IMO. The solution involves a quadratic Diophantine equation.

Example 1.44 (1988 IMO) *Let* a *and* b *be positive integers such that* $(1 + ab) | (a^2 + b^2)$. *Show that the integer* $(a^2 + b^2)/(1 + ab)$ *must be a perfect square.*

Solution. Suppose the proposition is false, so there are pairs of positive integers a, b such that $(a^2 + b^2)/(1 + ab)$ is a positive integer but not a square. From the set of counterexamples, pick one where $\max(a, b)$ is as small as possible. For this example, let $k = (a^2 + b^2)/(1 + ab)$. Clearly, $a \neq b$ since $a = b$ gives $k < 2$ and thus $k = 1$ (a square) as the only possibility. By symmetry, we may assume $a > b$. Observe that

$$a^2 + b^2 - k(1 + ab) = a^2 - kba + (b^2 - k) = 0$$

is a quadratic equation in a whose roots have sum kb and product $b^2 - k$. Thus, given the pair (a, b) satisfying $(a^2 + b^2)/(1 + ab) = k$, there is another such pair (a', b) where $a + a' = kb$ and $aa' = b^2 - k$. Since $(a')^2 + b^2 = k(a'b + 1)$ where k and b are positive integers, it is clear that $a' \geq 0$. Indeed, $a' > 0$ in view of the assumption that k is not a perfect square, in particular $k \neq b^2$. Finally,

$$a' = \frac{b^2 - k}{a} < \frac{b^2 - k}{b} < b$$

so $\max(a', b) = b < a = \max(a, b)$ and this violates our choice of (a, b) as a pair where $(a^2 + b^2)/(1 + ab)$ is a positive integer but not a square and where $\max(a, b)$ is as small as possible. □

There are several famous Diophantine equations whose solutions are known. Probably the most famous of these is the **Pythagorean equation**

$$x^2 + y^2 = z^2.$$

A **primitive solution** of this equation is one where x, y, and z have no common prime factor. In particular, in a primitive solution x or y must be odd. But x and y are not both odd since this would imply that $z^2 \equiv 2 \pmod 4$, which is impossible. Therefore we may assume that x is odd and y is even, so z is odd. With this understanding, we have the following famous result.

Theorem 1.15 *All positive primitive solutions of the equation* $x^2 + y^2 = z^2$ *where* y *is even are given by* $x = a^2 - b^2$, $y = 2ab$, $z = a^2 + b^2$, *where* a *and* b *are of opposite parity (one is even, the other odd),* $(a, b) = 1$ *and* $a > b > 0$.

To see this, we begin by writing the equation as

$$(z + x)(z - x) = y^2.$$

In view of the parity requirements on x, y, and z, we see that y, $z-x$, and $z + x$ are even. Thus we write

$$\frac{z + x}{2} \cdot \frac{z - x}{2} = \left(\frac{y}{2}\right)^2.$$

Since $(z, x) = 1$ it follows that $(z + x)/2$ and $(z-x)/2$ have no common prime factors, and the fact that their product is a perfect square yields

$$\frac{z + x}{2} = a^2 \quad \text{and} \quad \frac{z - x}{2} = b^2,$$

where $a > b > 0$ and $(a, b) = 1$. From these equations, it follows that $x = a^2 - b^2$, $y = 2ab$, and $z = a^2 + b^2$. Finally, since z is odd, a and b have opposite parity. □

Note. There is some evidence that the Babylonians knew these formulas for generating Pythagorean triples by 1600 B.C. In a Babylonian mathematical tablet known as *Plimpton 322*, there is a list of pairs of odd numbers

$$(x, z) = (119, 169), (3367, 4825), (4601, 6649), \ldots$$

These pairs correspond to the primitive Pythagorean triples generated by

$$(a, b) = (12, 5), (64, 27), (75, 32), \ldots$$

A homogeneous Diophantine equation like $x^2 + y^2 = z^2$ (x, y, z integers) can be phrased in terms of rational points on a curve.

Example 1.45 *Find all rational points (x, y) on the circle*

$$x^2 + y^2 = 1.$$

Solution. Pick a rational point on the curve, for example $x = -1, y = 0$. If (x, y) is another rational point on the curve, then the slope of the line through $(-1, 0)$ and (x, y) is rational. On the other hand, let t be an arbitrary rational number and consider the straight line $y = t(x + 1)$ that passes through $(-1, 0)$ and has slope t. This line intersects the circle at $(-1, 0)$ and again at the point (x, y) where

$$x^2 + t^2(x + 1)^2 = 1.$$

Solving this equation for x and then computing y we find

$$x = \frac{1 - t^2}{1 + t^2}, \qquad y = \frac{2t}{1 + t^2},$$

so that the point of intersection is rational. Thus we conclude that together with $(-1, 0)$, *all* rational points (x, y) that satisfy $x^2 + y^2 = 1$ are generated by the two equations above by letting t run through all possible rational values. □

Fermat's **method of infinite descent** is illustrated by the following celebrated result. This provided the first proved case of Fermat's Last Theorem.

Example 1.46 *Prove that the equation*

$$x^4 - y^4 = z^2$$

has no solution in positive integers.

Solution. Suppose that there are solutions of this equation in positive integers. If so, then there is a solution (x, y, z) where z takes the least possible value. Our goal is to obtain another solution (r, s, t) where $t < z$ and thus reach a contradiction. Our assumption that z is as small as possible clearly implies that $(x, y) = 1$ in the given solution, so (x^2, y^2, z) is a primitive Pythagorean triple. Hence we can write

$$x^2 = a^2 - b^2, \qquad y^2 = 2ab, \qquad z = a^2 + b^2,$$

where a and b have opposite parity, $(a, b) = 1$, and $a > b > 0$. Note further that a must be odd, since if a is even and b is odd then $x^2 = a^2 - b^2 \equiv -1 \pmod 4$, which is impossible. Thus a is odd, b is even and (x, b, a) is another primitive Pythagorean triple, so we can write

$$x = u^2 - v^2, \qquad b = 2uv, \qquad a = u^2 + v^2,$$

where u and v have opposite parity, $(u, v) = 1$, and $u > v > 0$. Note that

$$y^2 = 2ab = 4uv(u^2 + v^2),$$

and $(u, v) = (u, u^2 + v^2) = (v, u^2 + v^2) = 1$. It follows that u, v, and $u^2 + v^2$ are perfect squares:

$$u = r^2, \qquad v = s^2, \qquad u^2 + v^2 = t^2$$

But this yields

$$r^4 + s^4 = t^2,$$

where $t \le t^2 = u^2 + v^2 = a < a^2 + b^2 = z$, so we have reached our goal. This proves that the equation $x^4 + y^4 = z^2$ has no solution in positive integers and clearly takes care of the $n = 4$ case of Fermat's Last Theorem. □

If, instead of picking the smallest possible value for z in the first place, we had simply started with some assumed solution, the above argument would imply that there is an infinite sequence of solutions with descending values for "z," a clear impossibility. This way of thinking about it gives Fermat's method its name. Looking back at Example 1.42, we see that this kind of technique can be applied. Here the argument generates an infinite sequence of equations, namely $x^2 + y^2 + z^2 = 2^{2r}x^2y^2$, $(r = 0, 1, 2, \ldots)$ with a corresponding infinite sequence of positive solutions $(a, b, c), (a/2, b/2, c/2), (a/4, b/4, c/4), \ldots$, and this is clearly impossible.

Our final example of a famous Diophantine equation is one that is misnamed. It is referred to as **Pell's equation**, although Pell had practically nothing to do with it. The equation in question is

$$x^2 - Dy^2 = N,$$

where N and D are given integers and $D > 0$ is not a perfect square (so $\sqrt{D}$ is irrational).

In the special case $N = 1$ there is a quite satisfactory theory. The **fundamental solution** of the equation is the pair (x_1, y_1) which yields, out of all solutions (x, y), the smallest value exceeding 1 for the quantity $x + y\sqrt{D}$. It is known that a solution of $x^2 - Dy^2 = 1$ always exists, so there is always such a fundamental solution. The equation then has an infinite sequence of solutions (x_n, y_n) given by the formula

$$x_n + y_n\sqrt{D} = (x_1 + y_1\sqrt{D})^n.$$

It is easy to see that this formula yields solutions of the equation. Multiplication of this formula and its conjugate

$$x_n - y_n\sqrt{D} = (x_1 - y_1\sqrt{D})^n,$$

gives

$$x_n^2 - Dy_n^2 = (x_n - y_n\sqrt{D})(x_n + y_n\sqrt{D}) = (x_1^2 - Dy_1^2)^n = 1.$$

But it is also true that the above formula yields all solutions in natural numbers. A proof of this fact can be found in almost any text on number theory.

The more general equation

$$x^2 - Dy^2 = N$$

doesn't necessarily have a solution. However, if there is one solution of this equation, there are infinitely many of them.

Example 1.47 *Prove that $n^2 + (n+1)^2$ is a perfect square for infinitely many natural numbers n.*

Solution. We begin by writing the equation $n^2 + (n+1)^2 = m^2$ as

$$(2n+1)^2 - 2m^2 = -1.$$

Thus values of n for which $n^2 + (n+1)^2$ is a perfect square correspond to solutions of the Pell equation $x^2 - 2y^2 = -1$ where x is odd. One solution is $x = y = 1$. Let x_k and y_k be given by

$$x_k + y_k\sqrt{2} = (1+\sqrt{2})^{2k+1}, \quad k \geq 0.$$

Then

$$x_k^2 - 2y_k^2 = [(1-\sqrt{2})(1+\sqrt{2})]^{2k+1} = -1,$$

and by the binomial expansion $x_k = 1 + \binom{2k+1}{2}2 + \binom{2k+1}{4}2^2 + \cdots$ is odd. The first five values obtained this way are $n = 3$, $n = 20$, $n = 119$, $n = 696$, and $n = 4059$. □

In many cases of the equation $x^2 - Dy^2 = 1$, it is possible to spot the fundamental solution by inspection. For example, it is easy to see that $(2, 1)$ is the fundamental solution of

$$x^2 - 3y^2 = 1,$$

so all solutions in natural numbers are given by $x_n + y_n\sqrt{3} = (2 + \sqrt{3})^n$, $n \geq 1$. Thus the first few solutions are

$$(2,1),\ (7,4),\ (26,15),\ (97,56), \ldots$$

The sequence of solutions can be described effectively by means of a recurrence relation. Specifically, it is easy to show that $x_n =$

$4x_{n-1} - x_{n-2}$ and $y_n = 4y_{n-1} - y_{n-2}$ for $n \geq 3$. In the general case, the fundamental solution can be found from the continued fraction expansion of $\sqrt{D}$. If D is fairly large, the required computation can be formidable. There is a famous problem called the **cattle problem**, which is attributed to Archimedes. It is not quite clear what Archimedes had in mind, but by one interpretation the problem comes down to the equation

$$x^2 - 4729494\,y^2 = 1.$$

The fundamental solution of this equation is

$$x_1 = 109931986732829734979866232821433543901088049,$$

$$y_1 = 50549485234315033074477819735540408986340.$$

Exercises for Section 1.7

1. Solve the Diophantine epigram and find how long Diophantus lived.
2. Show that for every integer z there are integers x and y satisfying $x^2 - y^2 = z^3$. [1954 Putnam]
3. Find all solutions of $1 + x + x^2 + x^3 = 2^y$ in integers x, y.
4. Find all solutions of $x^2 + xy + y^2 = x^2y^2$ in integers x, y.
5. Show that the Diophantine equation $5m^2 - 6mn + 7n^2 = 1988$ has no solution.
6. Show that there are no integers a, b, c for which $a^2 + b^2 - 8c = 6$.
7. Determine all solutions in nonzero integers a and b of the equation $(a^2 + b)(a + b^2) = (a - b)^3$. [1987 USAMO]
8. Determine all solutions in nonnegative integers, if any, of the Diophantine equation

$$n_1^4 + n_2^4 + \cdots + n_{14}^4 = 1,599.$$

   (Two solutions in which $n_1, \ldots, n_{14}$ differ only by a permutation are considered to be the same.) [1979 USAMO]

9. Find all solutions of $1! + 2! + \cdots + x! = y^2$ in integers x, y.
10. Find all solutions in nonnegative integers a, b, c of the equation $3^a + 1 = 5^b + 7^c$.

11. Let n be an integer. Prove that if $2 + 2\sqrt{28n^2 + 1}$ is an integer, then it is a perfect square. [1969 József Kürschák Competition]

12. Let $p > 5$ be prime. Prove that the equation $x^4 + 4^x = p$ has no integer solution. [1977 József Kürschák Competition]

13. Find all integer solutions of the equation $x^4 + (x + 1)^4 = y^2 + (y + 1)^2$. [Arany Dániel Competition]

14. Find all rational numbers x such that $\sqrt{x^2 + x + 1}$ is also rational. *Hint:* Use the technique employed in Example 1.45.

15. Find all solutions in natural numbers m, n of the equation

$$m^2 = 1 + 2 + \cdots + n.$$

1.8 Quadratic Reciprocity

We say that an integer a satisfying $(a, m) = 1$ is a **quadratic residue** modulo m if there is a solution of the congruence $x^2 \equiv a \pmod{m}$. Otherwise, a is a **quadratic nonresidue** modulo m. Facts about quadratic residues can be used to solve many interesting problems, and there is a very beautiful theorem, stated by Legendre but first proved by Gauss, that is extremely useful in dealing with these problems. Gauss had conjectured the result on his own when he was 18; after some months of effort, he proved it. The theorem is called the **Law of Quadratic Reciprocity**. This was a favorite result of Gauss, and he returned to it several times over his career. In all, he found eight different proofs.

To state the law of quadratic reciprocity, we use a notation invented by Legendre. Let p be an odd prime. The **Legendre symbol** $\left(\frac{a}{p}\right)$ is defined as follows:

$$\left(\frac{a}{p}\right) = \begin{cases} 1 & \text{if } a \text{ is a quadratic residue modulo } p, \\ -1 & \text{if } a \text{ is a quadratic nonresidue modulo } p, \\ 0 & \text{if } p \mid a. \end{cases}$$

Theorem 1.16 (Law of Quadratic Reciprocity) *Let p and q be distinct odd primes. Then*

$$\left(\frac{p}{q}\right) = (-1)^{((p-1)/2)((q-1)/2)}.$$

Here is a brief sketch of a proof. The full details can be found in almost any text on number theory. First, using the fact that for every prime p there is a **primitive root**, that is to say an integer g such that the numbers $1, g, g^2, \ldots, g^{p-2}$ are pairwise incongruent modulo p, one can prove that

$$\left(\frac{a}{p}\right) \equiv a^{(p-1)/2} \pmod{p}.$$

The second step is the following lemma of Gauss: for any odd prime p and integer a satisfying $(a, p) = 1$, in the list of least positive residues of the integers $a, 2a, 3a, \ldots, (p-1)a/2$, there are n residues exceeding $p/2$, then $\left(\frac{a}{p}\right) = (-1)^n$. Using this fact, one can show that $(a, 2p) = 1$ implies

$$\left(\frac{a}{p}\right) = (-1)^r, \quad \text{where} \quad r = \sum_{j=1}^{(p-1)/2} \left\lfloor \frac{ja}{p} \right\rfloor.$$

Finally, by dividing the lattice points (x, y) satisfying $1 \leq x \leq (p-1)/2$ and $1 \leq y \leq (q-1)/2$ into two sets S_1 and S_2 according to whether $qx > py$ or $qx < py$, and counting the number of elements in each of the sets S_1 and S_2, it can be established that

$$\sum_{j=1}^{(p-1)/2} \left\lfloor \frac{jq}{p} \right\rfloor + \sum_{k=1}^{(q-1)/2} \left\lfloor \frac{kp}{q} \right\rfloor = \frac{p-1}{2}\frac{q-1}{2},$$

and this gives the law of quadratic reciprocity.

Another way to state the result is as follows: If either p or q is congruent to 1 modulo 4, then $\left(\frac{p}{q}\right) = \left(\frac{q}{p}\right)$; otherwise $\left(\frac{p}{q}\right) = -\left(\frac{q}{p}\right)$.

The following facts about the Legendre symbol are important.

Theorem 1.17 *Let p be an odd prime. Then*

(i) $a \equiv b \pmod{p} \Rightarrow \left(\frac{a}{p}\right) = \left(\frac{b}{p}\right)$,

(ii) $\left(\frac{a^2}{p}\right) = 1, \ a \not\equiv 0 \pmod p$,

(iii) $\left(\frac{a}{p}\right)\left(\frac{b}{p}\right) = \left(\frac{ab}{p}\right)$,

(iv) $\left(\frac{-1}{p}\right) = (-1)^{(p-1)/2}$,

(v) $\left(\frac{2}{p}\right) = (-1)^{(p^2-1)/8}$.

To illustrate how these results are used, let us determine whether 38 is a quadratic residue or quadratic nonresidue modulo 43. By repeated application of quadratic reciprocity and the results in Theorem 1.17, we have

$$\left(\frac{38}{43}\right) = \left(\frac{2}{43}\right)\left(\frac{19}{43}\right) = (-1)\cdot\left(-\left(\frac{43}{19}\right)\right)$$
$$= \left(\frac{5}{19}\right) = \left(\frac{19}{5}\right) = \left(\frac{-1}{5}\right) = 1.$$

(Our purpose is not to show the quickest way to find $\left(\frac{38}{43}\right)$ but to demonstrate quadratic reciprocity and the various results in Theorem 1.17.) For a quick proof, observe that $38 \equiv 9^2 \pmod{43}$.

Example 1.48 (1954 Putnam) *Prove that there are no integers x and y for which*

$$x^2 + 3xy - 2y^2 = 122.$$

Solution. Multiplying by 4 and completing the square, we obtain

$$(2x + 3y)^2 - 17y^2 = 488.$$

Since $488 \equiv 12 \pmod{17}$, it is enough to show that 12 is a quadratic nonresidue modulo 17. By the law of quadratic reciprocity,

$$\left(\frac{12}{17}\right) = \left(\frac{3}{17}\right)\left(\frac{4}{17}\right) = \left(\frac{3}{17}\right) = \left(\frac{17}{3}\right) = \left(\frac{2}{3}\right) = -1. \quad \square$$

Example 1.49 *(i) Prove that there are infinitely many primes of the form $3k - 1$. (ii) Prove that there are infinitely many primes of the form $3k + 1$.*

Solution. (i) To prove the first result, we follow the the lines of Euclid's proof that there are infinitely many primes. Suppose to the contrary that there are only finitely many primes congruent to -1 modulo 3 and let $p_1, p_2, \ldots, p_N$ be a complete list of such primes. Consider the number $M = 3(p_1 p_2 \cdots p_N) - 1$ and let q be one of its prime divisors. Then $q \neq 3$ and $q \neq p_1, p_2, \ldots, p_N$. This means that $q \equiv 1 \pmod 3$. But $M \equiv -1 \pmod 3$ and it is clearly impossible for each of its prime divisors to be congruent to 1 modulo 3. Thus we have a contradiction, meaning that our assumption that there are finitely many primes $\equiv -1 \pmod 3$ is false.

(ii) The second result requires a new trick, and this is where quadratic reciprocity comes in. Assume that $p_1, p_2, \ldots, p_N$ is a list of all primes congruent to 1 modulo 3, and set $M = 4(p_1 p_2 \cdots p_N)^2 + 3$. If q is a prime divisor of M then $q \neq 2, 3, p_1, p_2, \ldots, p_N$. This means that q is odd and $q \equiv -1 \pmod 3$. Since q divides $4(p_1 p_2 \cdots p_N)^2 + 3$, there is the additional requirement that -3 is a quadratic residue modulo q. But since

$$\left(\frac{-1}{q}\right) = (-1)^{(q-1)/2} \quad \text{and} \quad \left(\frac{3}{q}\right)\left(\frac{q}{3}\right) = (-1)^{(q-1)/2},$$

we have

$$\left(\frac{-3}{q}\right) = \left(\frac{-1}{q}\right)\left(\frac{3}{q}\right) = \left(\frac{q}{3}\right) = \left(\frac{2}{3}\right) = -1,$$

and thus the desired contradiction. □

Example 1.50 *Show that there are no integers a, b for which $2b^2 + 3$ divides $a^2 - 2$.*

Solution. Observe that $2b^2 + 3 \equiv 3 \pmod 8$ if b is even and $2b^2 + 3 \equiv 5 \pmod 8$ if b is odd. Thus $2b^2 + 3$ must have a prime divisor $p \equiv \pm 3 \pmod 8$ since every prime factor of $2b^2 + 3$ is congruent to ± 1, and this is clearly impossible. But $p \equiv \pm 3 \pmod 8$ yields $(2/p) = (-1)^{(p^2-1)/8} = -1$, so the given prime p does not divide $a^2 - 2$. Hence $2b^2 + 3$ does not divide $a^2 - 2$. □

Example 1.51 *Prove that if $m > 1$ and $n > 0$ have the same parity, then $3^n - 1$ is not divisible by $2^m - 1$.*

Solution. If m is even, then $2^m = 4^{m/2} \equiv 1 \pmod 3$ so $2^m - 1$ is divisible by 3. Hence $2^m - 1$ does not divide $3^n - 1$ for $n > 0$, regardless of the parity of n. Suppose m and n are both odd, $m > 1$, and $2^m - 1$ divides $3^n - 1$. Since $m > 1$ is odd, it follows that $2^m - 1 \equiv 7 \pmod{12}$.

(Note that $3|(2^{m-2}-2)$ so $12|(2^m-8)$.) Thus 2^m-1 has a prime divisor p such that $p \equiv 5 \pmod{12}$ or $p \equiv 7 \pmod{12}$. (An integer congruent to 7 modulo 12 is not divisible by 3, and it cannot be that each of its prime divisors is congruent to ± 1 modulo 12.) Since p divides 3^n-1 we have $3^{n+1} \equiv 3 \pmod p$, and since $n+1$ is even, this means that 3 is a quadratic residue modulo p. To obtain a contradiction and so complete the proof, we show that 3 is a quadratic nonresidue modulo p for $p \equiv 5 \pmod{12}$ and $p \equiv 7 \pmod{12}$. By quadratic reciprocity,

$$\left(\frac{3}{p}\right)\left(\frac{p}{3}\right) = (-1)^{(p-1)/2}.$$

Thus

$$p \equiv 5 \pmod{12} \Rightarrow \left(\frac{3}{p}\right) = \left(\frac{2}{3}\right) = -1$$

and

$$p \equiv 7 \pmod{12} \Rightarrow \left(\frac{3}{p}\right) = -\left(\frac{1}{3}\right) = -1.$$

Note that the condition of equal parity is essential. For example, $3^6-1 = 728$ is divisible by $2^3-1 = 7$. $\square$

By using properties of the Legendre symbol we can prove some striking facts about quadratic residues. These results have important applications in combinatorics. First let us note that for an odd prime p, there are $(p-1)/2$ quadratic residues modulo p and $(p-1)/2$ quadratic nonresidues. Perhaps the most elementary way to see this is to note first that we can generate the quadratic residues by squaring the numbers $1, 2, \ldots, p-1$ in turn. Then since $r^2 \equiv a^2 \pmod p$ implies that $r \equiv a \pmod p$ or $r \equiv -a \pmod p$ and since $r \not\equiv -r \pmod p$ for $r \not\equiv 0 \pmod p$, it follows by squaring $1, 2, \ldots, p-1$ in turn, that every quadratic residue occurs exactly two times. Thus there are $(p-1)/2$ quadratic residues modulo p. In terms of the Legendre symbol, this fact says

$$\sum_{r=0}^{p-1}\left(\frac{r}{p}\right) = 0. \tag{1.6}$$

Let r represent a nonzero residue class. Then by Euclid's Lemma there are integers x and y such that $rx + py = 1$. Thus $rx \equiv 1 \pmod p$

and in this case we write $x = r^{-1}$. In the following solution, we shall use the fact that for each such r there is a corresponding inverse. We shall also use the fact that for $a \not\equiv 0 \pmod{p}$, as r runs through the residue classes represented by $1, 2, \ldots, p-1$, the corresponding number $r^{-1}a$ does so as well, just in a different order.

Example 1.52 *For an odd prime p and a satisfying $0 < a < p$, let $N(a,p)$ denote the number of residues r satisfying*

$$\left(\frac{r}{p}\right) = \left(\frac{r+a}{p}\right) = +1.$$

Find a formula for $N(a,p)$.

Solution. First we evaluate the sum

$$S = \frac{1}{4}\sum_{r=0}^{p-1}\left(1 + \left(\frac{r}{p}\right)\right)\left(1 + \left(\frac{r+a}{p}\right)\right).$$

Multiplying out the summand and using (1.6) and other properties of the Legendre symbol, we obtain

$$S = \frac{1}{4}\left[p + \sum_{r=0}^{p-1}\left(\frac{r(r+a)}{p}\right)\right] = \frac{1}{4}\left[p + \sum_{r=1}^{p-1}\left(\frac{1+r^{-1}a}{p}\right)\right] = \frac{p-1}{4},$$

where the last step uses the fact that $1 + r^{-1}a$ goes through every nonzero value *except* 1 as r goes from 1 to $p-1$. Consider the contribution of the rth term to S. If both r and $r+a$ are quadratic residues modulo p then the contribution is 1, and if either r or $r+a$ is a quadratic nonresidue the contribution is 0. This is almost what we want. But there are two exceptional terms, namely $r = 0$ and $r = p - a$. Subtracting these two terms, we obtain

$$\begin{aligned} N(a,p) &= \frac{1}{4}\left[(p-1) - \left(1 + \left(\frac{a}{p}\right)\right) - \left(1 + \left(\frac{-a}{p}\right)\right)\right] \\ &= \frac{1}{4}\left[p - 3 - \left(\frac{a}{p}\right) - \left(\frac{-a}{p}\right)\right]. \end{aligned}$$

Specializing this formula, we have for $p \equiv 1 \pmod 4$,

$$N(a,p) = \begin{cases} \dfrac{p-5}{4}, & \left(\dfrac{a}{p}\right) = 1 \\[2ex] \dfrac{p-1}{4}, & \left(\dfrac{a}{p}\right) = -1, \end{cases}$$

and for $p \equiv 3 \pmod 4$,

$$N(a,p) = \frac{p-3}{4}. \quad \square$$

Olympiad Problems for Chapter 1

1. Prove that there are infinitely many natural numbers a with the following property: the number $z = n^4 + a$ is not prime for any natural number n. [1969 IMO]

2. Determine all three-digit numbers N having the property that N is divisible by 11, and $N/11$ is equal to the sum of the squares of the digits of N. [1960 IMO]

3. Let p and q be natural numbers such that

$$\frac{p}{q} = 1 - \frac{1}{2} + \frac{1}{3} - \cdots - \frac{1}{1318} + \frac{1}{1319}.$$

   Prove that p is divisible by 1979. [1979 IMO]

4. Let $\{a_n\}$ and $\{b_n\}$ be two sequences of integers defined as follows:

$$a_1 = a_2 = 1, \quad a_{n+1} = a_n + 2a_{n-1} \quad (n > 1),$$

$$b_1 = 1, b_2 = 7, \quad b_{n+1} = 2b_n + 3b_{n-1}, \quad (n > 1).$$

   Prove that there is no integer > 1 that occurs in both sequences. [1973 USAMO]

5. Prove that the set of integers of the form $2^k - 3$ $(k = 2, 3, \ldots)$ contains an infinite subset in which every two members are relatively prime. [1971 IMO]

6. Show that the cube roots of three distinct prime numbers cannot be three terms (not necessarily consecutive) of an arithmetic progression. [1973 USAMO]

7. Let $s(n)$ denote the sum of the digits of n when n is written in decimal. Evaluate

$$s(s(s(4444^{4444}))).$$

[1975 IMO]

8. Show that there is no natural number d that makes each of the numbers $2d-1$, $5d-1$, and $13d-1$ a perfect square. [1986 IMO]

9. Prove that for every integer $m > 2$ there exists an irrational number $r = r(m)$ such that $\lfloor r^k \rfloor \equiv -1 \pmod{m}$ for every natural number k. [Proposed for the 1987 IMO]

10. Prove that

$$\cos\frac{\pi}{7} - \cos\frac{2\pi}{7} + \cos\frac{3\pi}{7} = \frac{1}{2}.$$

[1963 IMO]

CHAPTER 2 Algebra

2.1 Basic Theorems and Techniques

The word **algebra** comes from the title *Hisâb al jabr w' al muquabalah* which the ninth century Arab mathematician al-Khowârizmî gave to his book on the solution of equations. (The mathematician's name is the etymological root of the word "algorithm.") The title of al-Khowârizmî's book translates to "science of reunion and opposition" and refers to the familiar processes of transposition and cancellation used in solving algebraic equations. Nowadays, "algebra" includes **classical** (mostly related to equations and inequalities involving polynomials), **linear** (vector space treatment of various problems, especially systems of linear equations) and **modern** or **abstract** (groups, rings, fields, and other basic algebraic structures). The emergence of modern algebra has come largely as a result of attempts to understand more clearly certain classical problems.

Our story begins with certain basic facts concerning polynomials. A **polynomial** in x over $\mathbb{C}$ (the field of complex numbers) is a formal expression

$$P(x) = c_0x^n + c_1x^{n-1} + \cdots + c_{n-1}x + c_n,$$

where $c_0, c_1, \ldots, c_n$ are complex numbers called the **coefficients** of the polynomial. With the understanding that the coefficient c_0 is nonzero, the **degree** of the polynomial, denoted by $\deg P$, is n. If all of the coefficients are zero, the polynomial is called the **zero polynomial** and its degree is not defined. The result obtained by substituting $a \in \mathbb{C}$ for the indeterminant x in $P(x)$ is denoted by $P(a)$. Polynomials in m variables take the form

$$P(x_1, x_2, \ldots, x_m) = \sum_{k_1, \ldots, k_m} c(k_1, \ldots, k_m)\, x_1^{k_1} x_2^{k_2} \cdots x_m^{k_m},$$

where $k_1, k_2, \ldots, k_m$ are nonnegative integers. Among the polynomials in more than one variable, the **symmetric polynomials** are of particular importance. A polynomial is **symmetric** if its value is unchanged whenever the variables are arbitrarily permuted. For example, the following is a symmetric polynomial in three variables:

$$P(a, b, c) = a(b + c)^2 + b(c + a)^2 + c(a + b)^2.$$

We shall have more to say concerning symmetric polynomials in the next section.

One very important fact concerning polynomials is the **Fundamental Theorem of Algebra**, which was discussed in §1.5. In addition to the Fundamental Theorem, the next three theorems are basic.

Theorem 2.1 (Uniqueness Theorem) *If $P(x)$ and $Q(x)$ are polynomials, each of degree at most n, and $P(x_i) = Q(x_i)$ for $i = 1, 2, \ldots, m$ where $x_1, x_2, \ldots, x_m$ are distinct complex numbers and $m > n$, then P and Q are identical.*

Theorem 2.2 (Division Algorithm) *If $F(x)$ and $G(x)$ are polynomials and $G(x)$ is not the zero polynomial, there exist unique polynomials $Q(x)$ and $R(x)$ such that*

$$F(x) = Q(x)G(x) + R(x),$$

where either (i) $R(x)$ is the zero polynomial or (ii) $\deg R < \deg G$.

Theorem 2.3 (Remainder and Factor Theorems) *Let $F(x)$ be a polynomial. When $F(x)$ is divided by $x - a$, the remainder is $F(a)$. Thus $x = a$ is a root of the equation $F(x) = 0$ if and only if $x - a$ is a factor of $F(x)$.*

The Remainder and Factor Theorems are easy consequences of the Division Algorithm. By setting $G(x) = x - a$ and applying the Division Algorithm, we see that $R(x)$ is constant regardless of whether (i) or (ii) holds. Substituting $x = a$ into $F(x) = Q(x)(x-a) + R(x)$, we see that the constant value of R is $F(a)$. It follows that if $F(a) = 0$ then $x - a$ is a factor of $F(x)$. Conversely, if $x - a$ is a factor of $F(x)$ then $F(a) = 0$.

Given

$$P(x) = c_0x^n + c_1x^{n-1} + \cdots + c_{n-1}x + c_n,$$

it is easy to find $Q(x)$ and $R = P(a)$ that satisfy $P(x) = Q(x)(x-a) + R$. Write

$$Q(x) = d_0x^{n-1} + d_1x^{n-2} + \cdots + d_{n-2}x + d_{n-1}$$

and compare coefficients in the equation

$$P(x) = Q(x)(x-a) + R.$$

Thus $c_0 = d_0$, $c_1 = d_1 - ad_0$, $\ldots$, $c_n = R - ad_{n-1}$. Transposing, we get the system

$$\begin{aligned} d_0 &= c_0, \\ d_1 &= c_1 + a\,d_0, \\ \vdots &= \vdots \\ d_{n-1} &= c_{n-1} + a\,d_{n-2}, \\ R &= c_n + a\,d_{n-1}. \end{aligned}$$

The process of using this system to compute recursively the coefficients of Q and the remainder R is known as **synthetic division**. Synthetic division is conveniently carried out in a "multiply and add" process performed in the tabular format shown below.

$$\begin{array}{cccccc} c_0 & c_1 & c_2 & \cdots & c_{n-1} & c_n \\ & ad_0 & ad_1 & \cdots & ad_{n-2} & ad_{n-1} \\ \hline d_0 & d_1 & d_2 & \cdots & d_{n-1} & R \end{array}$$

It is worth pausing here to note that the above equations clearly show that if P is a a polynomial with integral coefficients and a is an integer, then Q is also a polynomial with integral coefficients.

The following problems illustrate the Division Algorithm.

Example 2.1 *Find the remainder when $x^{81} + x^{49} + x^{25} + x^9 + x$ is divided by $x^3 - x$.*

Solution. One short solution is obtained by noting that the given polynomial can be written as

$$x(x^{80} - 1) + x(x^{48} - 1) + x(x^{24} - 1) + x(x^8 - 1) + 5x,$$

and that $x^3 - x = x(x^2 - 1)$ divides each of the first four terms on the right-hand side. Thus the remainder is $5x$. For the sake of variety, here is another solution. We are looking for the polynomial $R(x)$ of degree at most two that satisfies

$$x^{81} + x^{49} + x^{25} + x^9 + x = Q(x) \cdot (x^3 - x) + R(x)$$

for all x. Substituting $x = 0$, $x = 1$ and $x = -1$ in turn, we see that $R(0) = 0$, $R(1) = 5$ and $R(-1) = -5$. It is not hard to come up with a polynomial of degree at most two that satisfies these conditions; $R(x) = 5x$ clearly does the job. But the Uniqueness Theorem tells us that there is just one such polynomial. Hence the remainder is $R(x) = 5x$. □

Example 2.2 *Find the remainder when x^n is divided by $x^2 - x - 1$.*

Solution. We are looking for the unique polynomial R of degree at most one that satisfies

$$x^n = Q(x)(x^2 - x - 1) + R(x).$$

Let a and b be the roots of the equation $x^2 - x - 1 = 0$, namely $a = (1 + \sqrt{5})/2$ and $b = (1 - \sqrt{5})/2$. Substitution of $x = a$ and $x = b$ in $x^n = Q(x)(x^2 - x - 1) + R(x)$ yields $R(a) = a^n$ and $R(b) = b^n$, and the fact that $R(x)$ has degree at most one leads to

$$\begin{aligned} R(x) &= R(a) + \frac{R(a) - R(b)}{a - b}(x - a) \\ &= \frac{(a^n - b^n)x - ab(a^{n-1} - b^{n-1})}{a - b}. \end{aligned}$$

Using the relations $a - b = \sqrt{5}$, $ab = -1$ and the formula

$$F_n = \frac{a^n - b^n}{\sqrt{5}},$$

(Problem 10, **§1.2**), we recognize that the result can be expressed neatly in terms of Fibonacci numbers:

$$R(x) = F_n x + F_{n-1}.$$

It is worth noting that

$$Q(x) = F_1 x^{n-2} + F_2 x^{n-2} + \cdots + F_{n-1}.$$

In the set of exercises at the end of this section, there is a problem taken from the 1988 AIME which provides a good opportunity to use the result obtained in this example. □

Example 2.3 (Winter Competition, Romania, 1995) *Find all nonconstant polynomials P with real coefficients such that*

$$P(x^2) = P(x) \cdot P(x-1)$$

for all x.

Solution. Suppose $(x-a)$ is a factor of $P(x)$. Then $P(a^2) = P(a) \cdot P(a-1) = 0$. Continuing this way, we have

$$0 = P(a) = P(a^2) = P(a^4) = \cdots$$

But P is nonconstant, so not identically zero, and thus $P(x) = 0$ has a finite number of roots. It follows that two of the numbers $a, a^2, a^4, \ldots$ coincide and thus a is a root of unity. Consequently, $|a| = 1$. The equation $P(x^2) = P(x) \cdot P(x-1)$ also shows that

$$0 = P((a+1)^2) = P((a+1)^4) = P((a+1)^8) = \cdots$$

Thus, by the same argument as before, $|a+1| = 1$. Combining these facts, we have

$$(a+1)(\overline{a}+1) = (a+1)(1/a+1) = 1,$$

and this yields $a^2 + a + 1 = 0$ and thus $a = (-1 \pm i\sqrt{3})/2$. Since P has real coefficients, the fact that $(x-a)$ is a factor of $P(x)$ means that $x - \overline{a}$ is a factor as well; thus $(x-a)(x-\overline{a}) = x^2 + x + 1$ is a factor. No other nonconstant real factors are possible. Finally, $P(1) \neq 0$ implies $P(0) = 1$. Thus possible solutions are of the form $P(x) = (x^2+x+1)^n$ where n is a positive integer. It is easily checked that $P(x) = (x^2+x+1)^n$ does indeed satisfy $P(x^2) = P(x) \cdot P(x-1)$, so we have found all solutions. □

As we noted in the first chapter, if P is a polynomial with integral coefficients and a and b are artibrary integers, then $a - b$ divides $P(a) - P(b)$. This also follows immediately from Theorem 2.3. If P has integral coefficients and a is an integer, then

$$P(x) = (x - a)Q(x) + P(a),$$

where Q has integral coefficients. Substituting $x = b$, we draw the desired conclusion. The following problem uses a slight extension of this observation.

Example 2.4 (1974 IMO) *Let P be a nonconstant polynomial with integral coefficients. If $n(P)$ is the number of distinct integers k such that $(P(k))^2 = 1$, prove that $n(P) - \deg(P) \leq 2$, where $\deg(P)$ denotes the degree of the polyomial P.*

Solution. Let R_+ denote the set of integers k such that $P(k) = 1$. Similarly, let $R_- = \{k \in \mathbb{Z} | \ P(k) = -1\}$. We want to prove that $R = R_+ \cup R_-$ satisfies $|R| \leq \deg(P) + 2$. This conclusion is immediate if either R_+ or R_- is empty. In that case $|R| \leq \deg(P)$ since for a nonconstant polynomial P the equation $P(x) = 1$ (or $P(x) = -1$) has at most $\deg(P)$ distinct roots.

Now consider the case in which both R_+ and R_- are nonempty. We shall prove that in this case $|R| \leq 4$. This shows that $|R| \leq \deg(P) + 2$ whenever $\deg(P) \geq 2$. For $\deg(P) = 1$, it is enough to use $|R| \leq 2\deg(P) < \deg(P) + 2$. To prove that $|R| \leq 4$, we first claim that if $a \in R_+$, then either $R_- \subseteq \{a - 2, a - 1, a + 1\}$ or $R_- \subseteq \{a - 1, a + 1, a + 2\}$. To see this, suppose that $a \in R_+$ and that $b_1, b_2, \ldots, b_k \in R_-$. The Factor Theorem yields $P(x) + 1 = (x - b_1)\cdots(x - b_k)Q(x)$, where Q is a polynomial with integral coefficients. Substituting $P(a) = 1$, we obtain $(a - b_1)\cdots(a - b_k)Q(a) = 2$. Thus each of the factors $(a - b_1), \ldots, (a - b_k)$ must be ± 1 or ± 2, and $+2$ and -2 cannot both occur. This justifies our claim. In exactly the same way, if $b \in R_-$ then either $R_+ \subseteq \{b - 2, b - 1, b + 1\}$ or $R_+ \subseteq \{b - 1, b + 1, b + 2\}$. Suppose $R = \{c_1, c_2, \ldots, c_m\}$ where $a = c_1 < c_2 < \cdots < c_m$ and $m \geq 5$. Without loss of generality, we may assume $a \in R_+$. Then $R_- \subseteq \{a + 1, a + 2\}$ so $\{c_4, \ldots, c_m\} \subset R_+$. Since R_- is nonempty, we must assume either $c_2 \in R_-$ or $c_3 \in R_-$. But since $b \in R_-$ implies $R_+ \subseteq \{b - 2, b - 1, b + 1\}$

or $R_+ \subseteq \{b-1, b+1, b+2\}$, it is clear that both $c_2 \in R_-$ and $c_3 \in R_-$ are impossible. This contradiction shows that $|F| = m \leq 4$. □

The Factor Theorem points to the interplay between factoring and solving polynomial equations. The following examples illustrate this connection.

Example 2.5 *Factor* $(a + b + c)^3 - (a^3 + b^3 + c^3)$.

Solution. Denote the expression to be factored as $P(a, b, c)$ and observe that $P(a, -a, c) = c^3 - c^3 = 0$. By the Factor Theorem, $a + b$ is a factor of P. Similarly, $b + c$ and $c + a$ are factors. It follows that there is a constant K such that

$$P(a, b, c) = K(a + b)(b + c)(c - a)$$

for all a, b, c. To evaluate K, put $a = b = c = 1$. This yields $3^3 - 3 = 8K$ and hence $K = 3$. Thus

$$(a + b + c)^3 - (a^3 + b^3 + c^3) = 3(a + b)(b + c)(c + a). \quad \square$$

Example 2.6 *Factor completely*

$$a^3 + b^3 + c^3 - 3abc.$$

Solution. If $a + b + c = 0$ then

$$a^3 + b^3 + c^3 - 3abc = -(b + c)^3 + b^3 + c^3 + 3(b + c)bc = 0.$$

Thus $a + b + c$ is a factor. Now note that the value of $a^3 + b^3 + c^3 - 3abc$ is unchanged if b is replaced by ωb and c is replaced by $\omega^2 c$, where $\omega = (-1 + i\sqrt{3})/2$, so $\omega^3 = 1$ and $\omega^2 = \overline{\omega}$. Thus $a + \omega b + \omega^2 c$ is also a factor of $a^3 + b^3 - c^3 - 3abc$. By the same argument, $a + \omega^2 b + \omega c$ is a factor and thus the complete factorization is

$$a^3 + b^3 + c^3 - 3abc = (a + b + c)(a + \omega b + \overline{\omega} c)(a + \overline{\omega} b + \omega c). \quad \square$$

Example 2.7 (1976 USAMO) *If* $P(x)$, $Q(x)$ $R(x)$, *and* $S(x)$ *are polynomials such that*

$$P(x^5) + xQ(x^5) + x^2R(x^5) = (x^4 + x^3 + x^2 + x + 1)S(x),$$

prove that $x - 1$ *is a factor of* $P(x)$.

Solution. We observe that the right-hand side vanishes if x is any one of the complex fifth roots of unity. Let $\alpha = e^{2\pi i/5}$. Then substituting

$x = \alpha, \alpha^2, \alpha^3, \alpha^4$ in turn and using the fact that $\alpha^5 = 1$, we have

$$\begin{aligned}
P(1) + \alpha Q(1) + \alpha^2 R(1) &= 0, \\
P(1) + \alpha^2 Q(1) + \alpha^4 R(1) &= 0, \\
P(1) + \alpha^3 Q(1) + \alpha R(1) &= 0, \\
P(1) + \alpha^4 Q(1) + \alpha^3 R(1) &= 0.
\end{aligned}$$

Adding these equations and using $1 + \alpha + \alpha^2 + \alpha^3 + \alpha^4 = 0$, we obtain $4P(1) - Q(1) - R(1) = 0$. Multiplying the above equations by α, α^2, α^3, α^4, respectively, and then adding we find that $-P(1) - Q(1) - R(1) = 0$. Thus $4P(1) = Q(1) + R(1) = -P(1)$, so $P(1) = 0$ and $x - 1$ is a factor of $P(x)$. □

According to the Uniqueness Theorem, given $n + 1$ distinct complex numbers $a_0, a_1, \ldots, a_n$ and $n + 1$ arbitrary complex numbers $b_0, b_1, \ldots, b_n$, there is at most one polynomial P of degree n or less that satisfies $P(a_i) = b_i$ for $i = 0, 1, \ldots, n$. Such a polynomial exists; the process of constructing it is called **interpolation**. We shall discuss several interpolation problems, starting with some special examples and then presenting the general case.

Example 2.8 *Find the unique polynomial P of degree three that satisfies $P(0) = 0$ and $P(1) = P(2) = P(3) = 1$.*

Solution. According to the Factor Theorem, $P(x) - 1$ has factors $x - 1$, $x - 2$, and $x - 3$. It follows that there is a constant C such that

$$P(x) = 1 + C(x - 1)(x - 2)(x - 3).$$

To evaluate C, we use the fact that $P(0) = 0$. Thus

$$P(0) = 1 + C(-1)(-2)(-3) = 0,$$

from which we find $C = 1/6$. Thus the desired polynomial is

$$P(x) = 1 + \frac{1}{6}(x - 1)(x - 2)(x - 3). \quad \square$$

For the case where x is a real or complex number and k is a nonnegative integer, we define $\binom{x}{k}$ by

$$\binom{x}{k} = \frac{x(x - 1)\cdots(x - k + 1)}{k!}.$$

Then $\binom{x}{k}$ is a polynomial in x of degree k. If x is a nonnegative integer, the above definition yields the familiar binomial coefficient, namely the coefficient of $a^k b^{x-k}$ in the expansion of $(a+b)^x$.

Example 2.9 *Let $r \neq 0$ be given. Find the polynomial P of degree n or less that satisfies*

$$P(j) = r^j, \qquad j = 0, 1, \ldots, n.$$

Solution. We claim that

$$P(x) = \sum_{k=0}^{n} \binom{x}{k} (r-1)^k$$

is the desired polynomial. Clearly, P is a polynomial of degree at most n. Also, since the binomial coefficient $\binom{j}{k}$ vanishes for $k > j$, the Binomial Theorem yields

$$P(j) = \sum_{k=0}^{j} \binom{j}{k} (r-1)^k = r^j, \qquad j = 0, 1, \ldots, n.$$

But there is just one polynomial of degree at most n that satisfies $P(j) = r^j$ for $j = 0, 1, \ldots, n$. Thus our claim is justified. □

For future reference, note that

$$\begin{aligned} P(n+1) &= \sum_{k=0}^{n} \binom{n+1}{k} (r-1)^k \\ &= r^{n+1} - (r-1)^{n+1} \end{aligned} \tag{2.1}$$

Example 2.10 (1984 USAMO) *Suppose that P is a polynomial of degree $3n$ such that*

$$\begin{array}{lllllllll} P(0) & = & P(3) & = & \cdots & = & P(3n) & = & 2, \\ P(1) & = & P(4) & = & \cdots & = & P(3n-2) & = & 1, \\ P(2) & = & P(5) & = & \cdots & = & P(3n-1) & = & 0. \end{array}$$

If $P(3n+1) = 730$, determine n.

Solution. The fact that P has period three in the given range suggests that here is yet another problem where a complex cube root of unity could play a useful role. Indeed, note that

$$P(j) - 1 = \frac{2}{\sqrt{3}} \operatorname{Im} e^{\pi i/3} \omega^j, \qquad j = 0, 1, \ldots, 3n,$$

where $\omega = e^{2\pi i/3}$. We claim

$$P(x) = 1 + \frac{2}{\sqrt{3}} \sum_{k=0}^{3n} \binom{x}{k} \operatorname{Im} e^{\pi i/3}(\omega - 1)^k.$$

Just check; P is a polynomial of degree $3n$ and for $0 \le j \le 3n$,

$$\begin{aligned} P(j) &= 1 + \frac{2}{\sqrt{3}} \operatorname{Im} e^{\pi i/3} \sum_{k=0}^{j} \binom{j}{k} (\omega - 1)^k \\ &= 1 + \frac{2}{\sqrt{3}} \operatorname{Im} e^{\pi i/3} \omega^j. \end{aligned}$$

In view of (2.1),

$$\begin{aligned} P(3n+1) &= 1 + \frac{2}{\sqrt{3}} \operatorname{Im} e^{\pi i/3} \left[\omega^{3n+1} - (\omega - 1)^{3n+1}\right] \\ &= -\frac{2}{\sqrt{3}} \operatorname{Im} e^{\pi i/3} (\omega - 1)^{3n+1}. \end{aligned}$$

Since $\omega - 1 = i\sqrt{3}e^{\pi i/3}$, an application of De Moivre's theorem yields

$$P(3n+1) = 1 + 2(\sqrt{3})^{3n} \sin\left(\frac{(3n+1)\pi}{6}\right).$$

We are given that $P(3n+1) = 730 = 1 + 3^6$, and it is clear that this requires $n = 4$. □

Note: With a little more work, one can come up with a more efficient formula for computing $P(3n+1)$. Let $\lceil x \rceil$ denote the least integer $\ge x$. Then $P(3n+1) = 1 + (-1)^k 3^m$, where $k = n(n-1)/2$ and $m = \lceil 3n/2 \rceil$.

Our trick with the binomial expansion has worked very nicely in finding the polynomial that satisfies $P(j) = r^j$ for $j = 0, 1, \ldots, n$, but this situation is very special. We would like to have a general procedure for solving interpolation problems. Thus we seek an all-purpose formula for the unique polynomial P of degree n or less that satisfies

$$P(a_j) = b_j, \qquad j = 0, 1, \ldots, n.$$

(It is understood that $a_0, a_1, \ldots, a_n$ are *distinct* numbers.) Suppose that we can find polynomials $L_0, L_1, \ldots, L_n$, each of degree n or less, such that

$$L_k(a_j) = \begin{cases} 1 & \text{if } k = j, \\ 0 & \text{if } k \neq j. \end{cases}$$

Then our problem is solved, since

$$P(x) = \sum_{k=0}^{n} b_k L_k(x)$$

interpolates as desired. The construction of $L_k(x)$ follows the lines of our first example. If $L_k(a_j) = 0$ for all $j \neq k$, we know by the Factor Theorem that there is a constant C such that

$$L_k(x) = C(x - a_0)\cdots(x - a_{k-1})(x - a_{k+1})\cdots(x - a_n).$$

To evaluate C, substitute $x = a_k$ and use the condition $L_k(a_k) = 1$. Thus

$$1 = C(a_k - a_0)\cdots(a_k - a_{k-1})(a_k - a_{k+1})\cdots(a_k - a_n).$$

We can write the final result in compact form by using product symbol $\prod$. Thus

$$L_k(x) = \prod_{j \neq k} \left(\frac{x - a_j}{a_k - a_j} \right).$$

The formula

$$P(x) = \sum_{k=0}^{n} b_k L_k(x),$$

in which L_k is given by the preceding equation, is due to **Lagrange**. Another method for finding the interpolation polynomial, due to Newton, goes as follows. Write

$$\begin{aligned} P(x) = c_0 + c_1(x - a_0) + c_2(x - a_0)(x - a_1) + \cdots \\ + c_n(x - a_0)(x - a_1)\cdots(x - a_{n-1}), \end{aligned}$$

where $c_0, c_1, \ldots, c_n$ are constants to be determined. The form of Newton's polynomial allows one to determine the constants one at a time; thus

$$c_0 = b_0$$

$$c_1 = (b_1 - b_0)/(a_1 - a_0)$$
$$c_2 = \left(\frac{b_2 - b_0}{a_2 - a_0} - \frac{b_1 - b_0}{a_1 - a_0}\right)/(a_2 - a_1),$$

etc. The expressions on the right are called **divided differences**.

The problem of solving linear systems of equations is basic to the subject of **linear algebra**. In this problem, we are given an $m \times n$ array (**matrix**)

$$A = \begin{bmatrix} a_{11} & a_{12} & \cdots & a_{1n} \\ a_{21} & a_{22} & \cdots & a_{2n} \\ \vdots & & & \vdots \\ a_{m1} & a_{m2} & \cdots & a_{mn} \end{bmatrix},$$

an $m \times 1$ array (**vector**)

$$b = \begin{bmatrix} b_1 \\ b_2 \\ \vdots \\ b_m \end{bmatrix},$$

and asked to find all solutions, if any, of the system

$$\begin{array}{ccccccccc} a_{11}x_1 & + & a_{12}x_2 & + & \cdots & + & a_{1n}x_n & = & b_1 \\ a_{21}x_1 & + & a_{22}x_2 & + & \cdots & + & a_{2n}x_n & = & b_2 \\ \vdots & & \vdots & & & & \vdots & = & \vdots \\ a_{m1}x_1 & + & a_{m2}x_2 & + & \cdots & + & a_{mn}x_n & = & b_m. \end{array} \tag{2.2}$$

Here we briefly summarize the possibilities. By permuting equations and re-indexing the variables if necessary, we may assume that $a_{11} \neq 0$. We now divide the first equation by a_{11} and then in turn subtract an appropriate multiple of the first equation from each of the subsequent equations to eliminate x_1 from all but the first equation. The systematic continuation of this procedure is known as **Gaussian elimination**. At each stage of the procedure we might need to permute rows and re-index variables. The procedure stops only when it has been applied to all rows, or else every equation below a certain row has a vanishing left side. Thus the end result is an equivalent system (one with exactly the same solution set) of the form

$$\begin{array}{ccccccccccccc}
x_1 & + & c_{12}x_2 & + & \cdots & + & \cdots & + & c_{1n}x_n & = & d_1 \\
 & & x_2 & + & \cdots & + & \cdots & + & c_{2n}x_n & = & d_2 \\
 & & & & & & & & \vdots & = & \vdots \\
 & & & & x_r & + & \cdots & + & c_{rn}x_n & = & d_r \\
 & & & & & & & & 0 & = & d_{r+1} \\
 & & & & & & & & & \vdots & \\
 & & & & & & & & 0 & = & d_m.
\end{array}$$

This system is easy to analyze. First of all, if any one of the numbers $d_{r+1}, \ldots, d_m$ is not 0, then there is no solution. If $d_{r+1} = \cdots = d_m = 0$, then what happens depends on whether $r < n$ or $r = n$. If $r < n$, then $x_{r+1}, \ldots, x_n$ may be assigned arbitrary values. In this case, the system has infinitely many solutions. If $r = n$, then the system has a unique solution. A very significant fact is that the number r is a characteristic of the matrix A. Although there are choices to be made in the Gaussian elimination procedure that influence the particulars of the final system, the number of equations whose left side does not vanish, r, is independent of such choices. This number is called the **rank** of A.

From the preceding discussion, it is clear that the process of solving linear systems of equations is basically straightforward. Although possibly time-consuming, Gaussian elimination is sure-fire. However, before plunging in to a routine exercise in Gaussian elimination or turning the whole thing over to a computer, sometimes it pays to *think*. In each of the following examples of linear systems there is an observation that basically unlocks the problem and makes it unnecessary to wade through the Gaussian elimination procedure. We begin with a particularly easy example.

Example 2.11 (1986 AIME) *Determine the value of* $3x_4 + 2x_5$, *given that* x_1, x_2, x_3, x_4, x_5 *satisfy the system of equations given below:*

$$\begin{aligned} 2x_1 + x_2 + x_3 + x_4 + x_5 &= 6 \\ x_1 + 2x_2 + x_3 + x_4 + x_5 &= 12 \\ x_1 + x_2 + 2x_3 + x_4 + x_5 &= 24 \\ x_1 + x_2 + x_3 + 2x_4 + x_5 &= 48 \\ x_1 + x_2 + x_3 + x_4 + 2x_5 &= 96 \end{aligned}$$

Solution. Adding all five equations, we find

$$6(x_1 + x_2 + x_3 + x_4 + x_5) = 6(1 + 2 + 4 + 8 + 16),$$

so $x_1 + x_2 + x_3 + x_4 + x_5 = 31$. Subtraction of this equation from each of the given equations in turn yields

$$x_k = 6 \cdot 2^{k-1} - 31, \qquad k = 1, 2, \ldots, 5.$$

Thus $3x_4 + 2x_5 = 3 \cdot 17 + 2 \cdot 65 = 181$. □

Example 2.12 *Solve the system*

$$\begin{aligned} ax_1 + a^2x_2 + a^3x_3 + a^4x_4 &= 1 \\ bx_1 + b^2x_2 + b^3x_3 + b^4x_4 &= 1 \\ cx_1 + c^2x_2 + c^3x_3 + c^4x_4 &= 1 \\ dx_1 + d^2x_2 + d^3x_3 + d^4x_4 &= 1, \end{aligned}$$

where a, b, c, d *are distinct nonzero numbers.*

Solution. This is really an interpolation problem. If we let

$$P(t) = x_1t + x_2t^2 + x_3t^3 + x_4t^4,$$

then the given system states that the polynomial P satisfies $P(a) = P(b) = P(c) = P(d) = 1$. It is easy to construct P:

$$P(t) = 1 - \frac{(t-a)(t-b)(t-c)(t-d)}{abcd}.$$

Check: P is a polynomial of degree four with vanishing constant term ($P(0) = 0$) and $P(a) = P(b) = P(c) = P(d) = 1$. By the Uniqueness Theorem, there is just one such polynomial. Now we can just multiply out and read off the coefficients:

$$\begin{aligned} x_1 &= (abc + abd + acd + bcd)/(abcd), \\ x_2 &= -(ab + ac + ad + bc + bd + cd)/(abcd), \\ x_3 &= (a + b + c + d)/(abcd), \\ x_4 &= -1/(abcd). \quad \square \end{aligned}$$

In the solution to example 2.13 below, we will have occasion to use the method of **partial fractions**. Suppose that we given a **rational function**

$$R(x) = \frac{P(x)}{Q(x)},$$

where P and Q are polynomials,

$$Q(x) = (x - a_1)(x - a_2)\cdots(x - a_n),$$

and $\deg P < n$. (For simplicity, we assume that $a_1, a_2, \ldots, a_n$ are *distinct* complex numbers and that $P(a_k) \neq 0$ for $k = 1, 2, \ldots, n$.) The problem is to find $r_1, r_2, \ldots, r_n$ such that

$$R(x) = \sum_{k=1}^{n} \frac{r_k}{x - a_k}$$

for all $x \neq a_1, a_2, \ldots, a_n$. For such x it is legitimate to multiply both sides of the above equation by Q and then cancel common factors. Then we get the equation

$$\sum_{k=1}^{n} r_k \prod_{j \neq k} (x - a_j) = P(x),$$

in which both sides are polynomials. By the Uniqueness Theorem, the two polynomials are actually identical. (For identity, we just need the two polynomials to agree for $n + 1$ different values of x. Here we know much more; they agree for *every* x other than perhaps $a_1, \ldots, a_n$.) In view of the identity of the two polynomials, we may substitute any value we like. In particular, substituting $x = a_k$ allows us to solve for r_k:

$$r_k = \frac{P(a_k)}{\prod_{j \neq k}(a_k - a_j)}. \tag{2.3}$$

Example 2.13 *Solve the linear system (2.2) for the case where* $m = n$,

$$a_{ij} = \frac{1}{i+j-1}, \quad 1 \le i,j \le n,$$

$b_1 = b_2 = \cdots = b_{n-1} = 0$, *and* $b_n = 1$.

Solution. This is another case where interpolation provides a good weapon. Consider the rational function

$$R(t) = \sum_{k=1}^{n} \frac{x_k}{t+k}.$$

Then $R(t) = P(t)/Q(t)$ where P is a polynomial of degree $n-1$ or less and

$$Q(t) = (t+1)(t+2)\cdots(t+n).$$

The given system of equations yields

$$R(0) = R(1) = \cdots = R(n-2) = 0 \quad \text{and} \quad R(n-1) = 1.$$

In other words, $P(0) = \cdots = P(n-2) = 0$ and $P(n-1) = Q(n-1)$. Since P is of degree at most $n-1$, we know from the Uniqueness Theorem that P is determined by these conditions. Thus

$$P(t) = Ct(t-1)\cdots(t-n+2)$$

where C is determined by $P(n-1) = C(n-1)! = Q(n-1) = n(n+1)\cdots(2n-1)$. It follows that

$$C = \frac{(2n-1)(2n-2)\cdots(n)}{(n-1)!} = (2n-1)\binom{2n-2}{n-1}$$

and thus

$$R(t) = (2n-1)\binom{2n-2}{n-1}\frac{t(t-1)\cdots(t-n+2)}{(t+1)(t+2)\cdots(t+n)}.$$

Now we just need to determine $x_1, x_2, \ldots, x_n$ such that

$$\frac{x_1}{t+1} + \frac{x_2}{t+2} + \cdots + \frac{x_n}{t+n} = (2n-1)\binom{2n-2}{n-1}\frac{t(t-1)\cdots(t-n+2)}{(t+1)(t+2)\cdots(t+n)}.$$

This is the promised application of partial fractions. Using (2.3), we obtain

$$x_k = (2n-1)\binom{2n-2}{n-1}\frac{(-k)(-k-1)\cdots(-k-n+2)}{(-k+1)\cdots(-1)(1)\cdots(n-k)}$$

$$= (-1)^{n-k}(2n-1)\binom{2n-2}{n-1}\binom{n+k-2}{k-1}\binom{n-1}{k-1}. \qquad \square$$

A handy device for dealing with certain problems involving linear systems of equations is that of a **determinant.** Given a square matrix

$$A = \begin{bmatrix} a_{11} & a_{12} & \cdots & a_{1n} \\ a_{21} & a_{22} & \cdots & a_{2n} \\ \vdots & & & \vdots \\ a_{n1} & a_{n2} & \cdots & a_{nn} \end{bmatrix},$$

the corresponding determinant is

$$\det A = \sum_{\pi} \epsilon(\pi) \prod_{i=1}^{n} a_{i,\pi(i)}$$

where the sum is taken over all permutations π of $\{1, 2, \ldots, n\}$ and the **signature** $\epsilon(\pi)$ is $+1$ if π can be obtained by starting with the identity permutations $\{1, 2, \ldots, n\}$ and performing an even number of transpositions; otherwise $\epsilon(\pi) = -1$. There are many wonderful facts about determinants. We highlight two of them in the following theorem.

Theorem 2.4 *(i) If two rows (or columns) of A are proportional then* $\det A = 0$. *(ii) Let A_{ij} be $(-1)^{i+j}$ times the determinant of the matrix obtained by striking out row i and column j from A. For any i,*

$$\det A = \sum_{j=1}^{n} a_{ij}A_{ij}.$$

(Also $\det A = \sum_{i=1}^{n} a_{ij}A_{ij}$ *for any j.)*

For a detailed proof, see almost any book on linear algebra. The number A_{ij} is called the **cofactor** of a_{ij}, and the result in (ii) is called the **cofactor expansion** or **Laplace expansion.** Suppose $\det A \neq 0$ and we want to solve for x_i in the system (2.2) with $m = n$. We claim that

$$x_j = \frac{1}{\det A} \sum_{k=1}^{n} b_k A_{kj}, \qquad j = 1, 2, \ldots, n.$$

Just check:

$$\frac{1}{\det A}\sum_{j=1}^{n} a_{ij}\sum_{k=1}^{n} b_k A_{kj} = \frac{1}{\det A}\sum_{k=1}^{n} b_k \sum_{j=1}^{n} a_{ij}A_{kj} = b_i,$$

since for $i = k$ we have $\sum_{j=1}^{n} a_{ij}A_{kj} = \det A$ by the cofactor expansion, and $\sum_{j=1}^{n} a_{ij}A_{kj} = 0$ for $i \neq k$ since then the sum is (again by the cofactor expansion) the determinant of a matrix in which rows i and k are identical. The result just obtained in known as **Cramer's rule.**

Example 2.14 *Find the determinant of the* $n \times n$ **Hilbert matrix** H_n *whose* (i, j) *element is* $1/(i + j - 1)$.

Solution. This might be hard except for the result obtained in Example 2.13. Substituting $k = n$ in the final formula, we see that the value of x_n in the solution of the given $n \times n$ system is

$$x_n = (2n-1)\binom{2n-2}{n-1}^2.$$

But by applying Cramer's rule to the same system, we find that $x_n = \det H_{n-1}/\det H_n$. It follows that

$$\begin{aligned}\det H_n &= \det H_1 \frac{\det H_2}{\det H_1}\cdots\frac{\det H_n}{\det H_{n-1}}\\ &= \frac{1}{\prod_{k=1}^{n}(2k-1)\binom{2k-2}{k-1}^2}\\ &= \frac{\{1!2!\cdots(n-1)!\}^2}{1!2!\cdots(2n-1)!}. \quad \square\end{aligned}$$

Exercises for Section 2.1

1. Find the remainder when $x^{60} - 1$ is divided by $x^3 - 2$.
2. Find all integers c such that $x^2 - x + c$ is a factor of $x^{13} + x - 90$. [1963 Putnam]
3. Find all n such that $x^{n+1} - x^n + 1$ is divisible by $x^2 - x + 1$.
4. If $P(a) = a$, $P(b) = b$, and $P(c) = c$ where a, b, and c are distinct numbers, what is the remainder when the polynomial P is divided by $(x - a)(x - b)(x - c)$?

5. Determine k such that

$$x^5 + y^5 + z^5 + k(x^2 + y^2 + z^2)(x^3 + y^3 + z^3)$$

has $x + y + z$ as a factor.

6. Find a if a and b are integers such that $x^2 - x - 1$ is a factor of $ax^{17} + bx^{16} + 1$. [1988 AIME]

7. Determine all pairs of positive integers (m, n) such that $1 + x^n + x^{2n} + \cdots + x^{mn}$ is divisible by $1 + x + x^2 + \cdots + x^m$. [1977 USAMO]

8. Factor $a(b - c)^3 + b(c - a)^3 + c(a - b)^3$.

9. Factor $(a + b)^7 - a^7 - b^7$.

10. Factor $8(a + b + c)^3 - (a + b)^3 - (b + c)^3 - (c + a)^3$.

11. Find the unique polynomial P of degree n that satisfies

$$P(j) = \frac{1}{j + 1}, \qquad j = 0, 1, \ldots, n.$$

Hint: Consider $(x + 1)P(x) - 1$.

12. If P is the polynomial of degree n or less that satisfies $P(k) = \sin(k\pi/n)$ for $k = 0, 1, \ldots, n$, what is $P(n + 1)$?

13. A polynomial P of degree 990 satisfies $P(k) = F_k$ for $k = 992, 993, \ldots, 1982$, where F_k denotes the kth Fibonacci number. Prove that $P(1983) = F_{1983} - 1$. [Proposed for the 1983 IMO]

14. Find all triples of nonnegative real numbers (x, y, z) satisfying

$$\begin{aligned} \sqrt[3]{x} - \sqrt[3]{y} - \sqrt[3]{z} &= 16 \\ \sqrt[4]{x} - \sqrt[4]{y} - \sqrt[4]{z} &= 8 \\ \sqrt[6]{x} - \sqrt[6]{y} - \sqrt[6]{z} &= 4. \end{aligned}$$

15. If $x_1, x_2, \ldots, x_n$ satisfy

$$\sum_{j=1}^{n} \frac{x_j}{i + j} = \frac{4}{2i + 1}, \qquad i = 1, 2, \ldots, n,$$

determine

$$\sum_{j=1}^{n} \frac{x_j}{2j + 1}.$$

Hint: Let

$$R(t) = \sum_{j=1}^{n} \frac{x_j}{t + j},$$

and note that the desired sum is simply $\frac{1}{2}R(\frac{1}{2})$. Solve the interpolation problem that yields $R(t)$.

Note: This problem appeared in the *American Mathematical Monthly* several years ago. It was posed by the famous mathematical physicist Freeman Dyson. The problem arose in an investigation of the diffraction of light by another famous physicist, Julian Schwinger. (Along with Richard Feynman and Sinichiro Tomonaga, Schwinger won the 1965 Nobel Prize in Physics for his work on quantum electrodynamics.)

2.2 Polynomial Equations

2.2.1 Solving Polynomial Equations

Polynomial equations have played an important role in the development of mathematics. By 2000 B.C., Babylonian mathematicians had developed the method of solving quadratic equations by completing the square. However, a general method for solving cubic andquartic equations was not discovered until the sixteenth century. Then Scipione del Ferro and Niccolo Fontana of Brescia (Tartaglia) found the general solution of the cubic and Lodovico Ferrari developed a method of solving an arbitrary quartic equation. Finally, in the early nineteenth century, Niels Henrik Abel proved that there is no general method for solving polynomial equations of degree five (or higher) using rational operations and extraction of radicals, and Évariste Galois found a complete theory of solvability. The theory of Galois involves deep connections between **groups** and **fields** and is one of the most impressive abstract theories in all of mathematics.

We begin with the basic problem: find all roots of the polynomial equation

$$P(x) = c_0x^n + c_1x^{n-1} + \cdots + c_n = 0.$$

In general, $c_0, c_1, \ldots, c_n$ are complex numbers, and we know from the Fundamental Theorem of Algebra that there are complex numbers x that satisfy the equation. A root of the equation $P(x) = 0$ is also referred to as a **zero** of the polynomial P. The roots can be approximated in various ways, and there are efficient computer algorithms for constructing such approximations. However, to **solve** the equation means something quite specific. It means to specify all of the roots of the equation exactly by a finite number of standard operations ($+$, $-$, $\times$, $\div$, $\sqrt[n]{\ }$) applied to the coefficients. Everyone is familiar with how this works for the quadratic equation

$$ax^2 + bx + c = 0.$$

By the Babylonian technique of completing the square, one finds the complete solution set,

$$S = \left\{ \frac{-b \pm \sqrt{b^2 - 4ac}}{2a} \right\}.$$

As we shall see, such formulas also exist for cubic and quartic equations. But, as Abel first proved, arbitrary equations of degree five and higher cannot be solved in such a way.

Since polynomial equations cannot be solved in general, we may view a solvable equation as an example of divine providence. It is our job to take advantage of this good fortune and to obtain the solution in an intelligent fashion. The following plan of attack is suggested:

- Rational roots?
- Special form?
- If all else fails...

Recall the Rational Root Theorem: If $c_0, c_1, \ldots, c_n$ are integers and $x = a/b$ is a rational root of the equation, then $a|c_n$ and $b|c_0$. This is a very powerful fact, and it should be used at the outset whenever it applies.

Example 2.15 *Solve the equation* $3x^3 - 7x^2 + 17x - 5 = 0$.

Solution. We first look for rational solutions. From the Rational Root Theorem, if $x = a/b$ is a root then $a|5$ and $b|3$. Trying the possible combinations, we find that $x = 1/3$ is in fact a root. Now we can factor:

$$(3x-1)(x^2-2x+5)=0.$$

Thus the solution set is $S = \{1/3,\ 1 \pm 2i\}$. □

One kind of special form involves symmetry of the coefficients.

Example 2.16 *Solve the quartic equation*

$$x^4+2ax^3+bx^2+2ax+1=0.$$

Solution. Noting that $(x^2+ax+1)^2 = x^4+2ax^3+(a^2+2)x^2+2ax+1$, we see that the equation can be written as

$$(x^2+ax+1)^2-(a^2+2-b)x^2=0.$$

The left-hand side is a difference of squares, so we can factor and thus obtain

$$(x^2+(a+\sqrt{a^2+2-b})x+1)(x^2+(a-\sqrt{a^2+2-b})x+1)=0.$$

Now the problem is reduced to solving two quadratic equations. We leave the details to the reader. □

The equation $x^4+2ax^3+bx^2+2ax+1=0$ is a special case of a **reciprocal equation**. An equation of even degree

$$c_0x^{2n}+c_1x^{2n-1}+\cdots+c_{2n}=0$$

is reciprocal if $c_k = c_{2n-k}$ for all k; in other words, the sequence of coefficients reads the same from right to left as it does from left to right. For such an equation, the transformation

$$z=x+x^{-1}$$

reduces the problem to that of solving a polynomial equation of degree n (half the original degree). Applying this technique to our quartic equation in Example 2.16, we first divide by x^2 to obtain

$$x^2+2ax+b+2ax^{-1}+x^{-2}=0,$$

and then make the substitution $z = x+x^{-1}$ and use the fact that $x^2+x^{-2}=z^2-2$ to obtain

$$z^2+2az+(b-2)=0.$$

Thus $z = -a \pm \sqrt{a^2+2-b}$, and the solution is completed as before.

There are many other cases where the equation is special and can be solved easily if one makes the right observation.

Example 2.17 *Solve* $x^4 + 2x^3 + 7x^2 + 6x + 8 = 0$.

Solution. Observe that the equation can be written as

$$(x^2 + x)^2 + 6(x^2 + x) + 8 = 0.$$

Thus

$$(x^2 + x + 4)(x^2 + x + 2) = 0,$$

and the complete solution set is

$$S = \left\{ \frac{-1 \pm i\sqrt{15}}{2}, \frac{-1 \pm i\sqrt{7}}{2} \right\}. \quad \square$$

Special polynomial equations include those whose roots are special values of trigonometric functions. The **Chebyshev polynomial** of degree n is $T_n(x) = \cos(n \cos^{-1} x)$. The polynomials $T_0(x), T_1(x), T_2(x), \ldots$ satisfy

$$T_{n+1}(x) = 2xT_n(x) - T_{n-1}(x), \quad T_0(x) = 1, \ T_1(x) = x.$$

In view of the defining expression for $T_n(x)$, the roots of the equation $T_n(x) = 0$ are given by

$$x = \cos\left(\frac{(2k-1)\pi}{2n}\right), \quad k = 1, 2, \ldots, n.$$

Example 2.18 (Proposed for the 1991 IMO) *Show that the zeros of* $x^8 - 92x^6 + 134x^4 - 28x^2 + 1$ *are*

$$x = \tan(r\pi/15), \quad 1 \le r < 15 \textit{ and } GCD(r, 15) = 1.$$

Solution. It suffices to check that the unique monic polynomial whose zeros are the eight numbers specified is $x^8 - 92x^6 + 134x^4 - 28x^2 + 1$. Note that if $\theta = r\pi/15$ for some $1 \le r < 15$ satisfying $\text{GCD}(r, 15) = 1$, then

$$\tan^2(5\theta) = 3 \quad \text{and} \quad \tan^2\theta \ne 3. \tag{2.4}$$

Moreover, the values $\tan(r\pi/15)$ for $r = 1, 2, 4, 7, 8, 11, 13, 14$ include all distinct values of $\tan\theta$ satisfying (2.4). Thus it suffices to find the equation whose roots are $x = \tan\theta$ where (2.4) holds. If $x = \tan\theta$ then

$$\tan(5\theta) = \frac{\operatorname{Im}(1+ix)^5}{\operatorname{Re}(1+ix)^5} = \frac{x(x^4-10x^2+5)}{5x^4-10x^2+1},$$

so $\tan^2(5\theta) = 3$ becomes

$$\frac{x^2(x^4-10x^2+5)^2}{(5x^4-10x^2+1)^2} = 3.$$

This reduces to

$$x^2(x^8-20x^6+110x^4-100x^2+25) = 3(25x^8-100x^6+110x^4-20x^2+1),$$

and so

$$x^{10}-95x^8+410x^6-430x^4+85x^2-3 = 0.$$

Using synthetic division to divide by x^2-3, we obtain

1	−95	410	−430	85	−3
	3	−276	402	−84	3
1	−92	134	−28	1	

so

$$x^8-92x^6+134x^4-28x^2+1$$

is the desired polynomial. □

Finally we turn to the last alternative ("if all else fails"). If the equation is of degree at most four and is not otherwise special, we can apply the methods of Tartaglia and Ferrari to find the solution. The general cubic equation,

$$c_0x^3+c_1x^2+c_2x+c_3 = 0,$$

can be transformed (by dividing by c_0 and letting $z = x + \frac{c_1}{3c_0}$) to an equation of the form $z^3+pz+q = 0$. Thus, we may assume this form at the outset and take the given equation to be

$$x^3+px+q = 0.$$

To solve this equation, we substitute $x = u+v$ to obtain

$$u^3+v^3+(u+v)(3uv+p)+q = 0,$$

and note that we are free to require $uv = -p/3$. Then u^3 and v^3 are the roots of the equation $z^2 + qz - p^3/27 = 0$. Solving this equation, we obtain

$$u^3,\ v^3 = -\frac{q}{2} \pm \sqrt{R},$$

where

$$R = \left(\frac{q}{2}\right)^2 + \left(\frac{p}{3}\right)^3.$$

Now we may choose cube roots so that

$$A = \sqrt[3]{-\frac{q}{2} + \sqrt{R}} \quad \text{and} \quad B = \sqrt[3]{-\frac{q}{2} - \sqrt{R}}$$

give $AB = -p/3$. Then $u = A$, $v = B$ satisfy our requirements, and thus $x = A + B$ is one solution of the equation. Let $\omega = (-1 + i\sqrt{3})/2$ and $\overline{\omega} = \omega^2 = (-1 - i\sqrt{3})/2$ denote the complex cube roots of unity. The remaining solutions are obtained by letting $u = \omega A$ or $u = \overline{\omega} A$ and choosing v so that $uv = -p/3$. Thus

$$S = \{A + B,\ \omega A + \overline{\omega} B,\ \overline{\omega} A + \omega B\}$$

is the complete solution set of the equation.

Example 2.19 *Solve* $x^3 - 3x - 1 = 0$.

Solution. Setting $x = u + v$ where $uv = 1$, we find that

$$u^6 + u^3 + 1 = 0.$$

Thus $u^3 = (-1 \pm i\sqrt{3})/2 = e^{\pm 2\pi i/3}$, and it follows that $u = A$, $v = B$ yields a solution, where

$$A = \cos(2\pi/9) + i\sin(2\pi/9) \quad \text{and} \quad B = \cos(2\pi/9) - i\sin(2\pi/9).$$

Thus $x = A + B = 2\cos(2\pi/9)$ is one root of the equation $x^3 - 3x + 1 = 0$. Putting the other two roots $(\omega A + \overline{\omega} B,\ \overline{\omega} A + \omega B)$ in trigonometric form, we find that the complete solution set of $x^3 - 3x + 1 = 0$ is

$$S = \{2\cos(2\pi/9),\ 2\cos(4\pi/9),\ 2\cos(8\pi/9)\}. \quad \square$$

The general quartic equation can be solved in several different ways. In each method of solution, the problem is reduced to solving a cubic equation called the **resolvent**. In Ferrari's method, one takes the quartic equation to be

$$x^4 + 2ax^3 + bx^2 + 2cx + d = 0.$$

Transposing to obtain

$$x^4 + 2ax^3 = -bx^2 - 2cx - d$$

and then adding $2rx^2 + (ax + r)^2$ to both sides makes the left-hand side equal to $(x^2 + ax + r)^2$. If r can be chosen to make the right-hand side a perfect square, then the way is clear to a solution. The right-hand side is

$$(2r + a^2 - b)x^2 + 2(ar - c)x + (r^2 - d),$$

and in order to make this a perfect square, we require the discriminant to be zero. Thus we require

$$(ar - c)^2 - (2r + a^2 - b)(r^2 - d) = 0,$$

or

$$2r^3 - br^2 + 2(ac - d)r + (bd - a^2d - c^2) = 0.$$

This is the cubic resolvent. Assuming a, b, c, d to be real, there is always a real number r that satisfies the resolvent equation.

Example 2.20 *Solve* $x^4 - 26x^2 + 72x - 11 = 0.$

Solution. In this example, $a = 0,\ b = -26,\ c = 36$, and $d = -11$. The cubic resolvent,

$$(-36)^2 - (2r + 26)(r^2 + 11) = 0,$$

is clearly satisfied by $r = 5$. Adding $10x^2 + 25$ to each side of the equation

$$x^4 = 26x^2 - 72x + 11,$$

we obtain

$$(x^2 + 5)^2 = (6x - 6)^2.$$

Thus the quartic equation factors as

$$(x^2 - 6x + 11)(x^2 + 6x - 1) = 0,$$

and the complete solution set is $S = \{3 \pm i\sqrt{2},\ -3 \pm \sqrt{10}\}$. □

Occasionally we need to solve a polynomial equation containing one or more **parameters**. The roots of the equation are then functions of these parameters. Even in the simplest of cases, these functions can be fairly complicated. In the following examples, we shall take the parameter (denoted by a) to be real.

Example 2.21 *Solve the equation* $(2a-1)x^2 - ax + 1 - a = 0$.

Solution. Factoring, we obtain

$$(x-1)\{(2a-1)x + a - 1\} = 0.$$

Hence $x = 1$ is a solution, and if $a \neq \frac{1}{2}$, then $x = (1-a)/(2a-1)$ is also a solution. (If $a = \frac{1}{2}$, then $x = 1$ is the only solution.) □

Example 2.22 *Solve the equation* $z^4 - 2z^3 + z^2 - a = 0$ *and find values of* a *for which all roots are real.*

Solution. Note that we can write the equation as $\{z(z-1)\}^2 = a$. For $a \geq 0$, we have $z(z-1) = \pm\sqrt{a}$ and thus find the solution set to be

$$S = \left\{ \frac{1 \pm \sqrt{1 \pm 4\sqrt{a}}}{2} \right\}.$$

It is apparent that all roots are real if and only if $0 \leq a \leq 1/16$. For completeness, let us indicate what happens in the other cases. Let $z = x + iy$. For $a < 0$, the roots are two complex conjugate pairs. (It is not hard to show that in this case all four roots lie on the hyperbola $(x - \frac{1}{2})^2 - y^2 = \frac{1}{4}$.) For $a > 1/16$, two of the roots are real and the other two satisfy $x = 1/2$, $y \neq 0$. It is instructive to sketch the trajectories of the roots in the complex plane as a varies. □

Problems involving complex roots of polynomial equations with real parameters are important in applied mathematics. In the study of stability of physical systems, it is important to know the values of the parameter(s) for which all roots have negative real part.

2.2.2 Symmetric Functions

Now we come to a surprising and beautiful fact. Although there is no general formula that takes us from the coefficients $c_0, c_1, \ldots, c_n$

of the polynomial equation

$$c_0x^n + c_1x^{n-1} + \cdots + c_n = 0$$

to its roots $x_1, x_2, \ldots, x_n$, there are formulas that take us from $c_0, c_1, \ldots, c_n$ to a large and important class of **functions** of the roots. These are the **symmetric functions**. A symmetric function of $x_1, x_2, \ldots, x_n$ is one whose value is unchanged if $x_1, x_2, \ldots, x_n$ are permuted arbitrarily. For example, each of the following is a symmetric function of three variables:

$$\begin{aligned} P(x_1, x_2, x_3) &= x_1x_2 + x_2x_3 + x_3x_1, \\ Q(x_1, x_2, x_3) &= x_1^3 + x_2^3 + x_3^3, \\ R(x_1, x_2, x_3) &= \frac{x_2 + x_3}{x_1} + \frac{x_3 + x_1}{x_2} + \frac{x_1 + x_2}{x_3}. \end{aligned}$$

Certain symmetric functions serve as building blocks for all the rest. Let

$$\sigma_k = \sum x_{i_1}x_{i_2}\cdots x_{i_k},$$

where the sum is taken over all $\binom{n}{k}$ choices of the indices $i_1, i_2, \ldots, i_k$ from $\{1, 2, \ldots, n\}$. Then σ_k is called the kth **elementary symmetric function** of $x_1, x_2, \ldots, x_n$.

Theorem 2.5 (Symmetric Function Theorem) *Every symmetric polynomial function of $x_1, x_2, \ldots, x_n$ is a polynomial function of $\sigma_1, \sigma_2, \ldots, \sigma_n$. The same conclusion holds if "polynomial" is replaced by "rational function."*

As an illustration, for $n = 3$ the elementary symmetric functions are

$$\begin{aligned} \sigma_1 &= x_1 + x_2 + x_3, \\ \sigma_2 &= x_1x_2 + x_2x_3 + x_3x_1, \\ \sigma_3 &= x_1x_2x_3, \end{aligned}$$

and it is easy to check that the examples given earlier can be expressed in terms of these as follows:

$$\begin{aligned} x_1x_2 + x_2x_3 + x_3x_1 &= \sigma_2, \\ x_1^3 + x_2^3 + x_3^3 &= \sigma_1^3 - 3\sigma_1\sigma_2 + 3\sigma_3, \end{aligned}$$

$$\frac{x_2 + x_3}{x_1} + \frac{x_1 + x_3}{x_2} + \frac{x_2 + x_1}{x_3} = \frac{\sigma_1\sigma_2 - 3\sigma_3}{\sigma_3}.$$

The Symmetric Function Theorem provides us with a very important piece of knowledge. Another important fact involves the relationship between the coefficients of a polynomial and the elementary symmetric functions of its zeros.

Theorem 2.6 *Let $x_1, x_2, \ldots, x_n$ be the roots of the polynomial equation*

$$x^n + c_1x^{n-1} + \cdots + c_n = 0,$$

and let σ_k be the kth elementary symmetric function of the x_i. Then

$$\sigma_k = (-1)^k c_k, \qquad k = 1, 2, \ldots, n.$$

This result is transparent. On the left-hand side of the equation

$$x^n + c_1x^{n-1} + \cdots + c_n = (x - x_1)(x - x_2)\cdots(x - x_n)$$

the coefficient of x^{n-k} is c_k; on the right-hand side, the coefficient of x^{n-k} is $(-1)^k$ times the sum of all $\binom{n}{k}$ products of k of the x_i. Thus $c_k = (-1)^k\sigma_k$. The last two theorems have many applications.

Example 2.23 *Find all solutions of the system of equations*

$$\begin{aligned} x + y + z &= 0, \\ x^2 + y^2 + z^2 &= 6ab, \\ x^3 + y^3 + z^3 &= 3(a^3 + b^3). \end{aligned}$$

Solution. The key is that the left side of each equation is a symmetric function of x, y, z. This suggests that we can use the information given to construct a polynomial equation whose roots are x, y, z. Let

$$P(t) = (t - x)(t - y)(t - z) = t^3 + c_1t^2 + c_2t + c_3.$$

Then $c_1 = -(x + y + z) = 0$ and

$$c_2 = xy + yz + zx = \frac{(x + y + z)^2 - (x^2 + y^2 + z^2)}{2} = -3ab.$$

Finally, $P(x) = P(y) = P(z) = 0$ yields

$$(x^3 + y^3 + z^3) + c_1(x^2 + y^2 + z^2) + c_2(x + y + z) + 3c_3 = 0,$$

from which we find

$$c_3 = -\frac{x^3 + y^3 + z^3}{3} = -(a^3 + b^3).$$

Thus x, y, z are the roots of the cubic equation

$$t^3 - 3abt - (a^3 + b^3) = 0.$$

Observe that $t = a + b$ is one of the roots. Now we can factor to obtain

$$\{t - (a + b)\}\{t^2 + (a + b)t + (a^2 - ab + b^2)\},$$

and so find the complete solution set:

$$S = \{a + b,\ a\omega + b\overline{\omega},\ a\overline{\omega} + b\omega\}. \quad \square$$

(As before, ω and $\overline{\omega} = \omega^2$ are the two complex cube roots of unity.)

An important class of symmetric functions are the power sums

$$S_p = x_1^p + x_2^p + \cdots + x_n^p,$$

where p is a nonnegative integer. The power sums are easily computed in terms of the numbers $c_k = (-1)^k \sigma_k$, $k = 1, 2, \ldots, n$ using a beautiful system of equations discovered by Newton.

Theorem 2.7 (Newton's Formulas for Power Sums) *Let*

$$S_p = x_1^p + x_2^p + \cdots + x_n^p,$$

where $x_1, x_2, \ldots, x_n$ are the roots of

$$x^n + c_1 x^{n-1} + \cdots + c_n = 0.$$

Then

$$\begin{array}{rcrcrcrcrcl}
 & & & & & & S_1 & + & c_1 & = & 0, \\
 & & & & S_2 & + & c_1 S_1 & + & 2c_2 & = & 0, \\
 & & S_3 & + & c_1 S_2 & + & c_2 S_1 & + & 3c_3 & = & 0, \\
 & & \vdots & & \vdots & & \vdots & & \vdots & & \vdots \\
S_n & + & c_1 S_{n-1} & + & \cdots & + & c_{n-1} S_1 & + & nc_n & = & 0,
\end{array}$$

and

$$S_p + c_1 S_{p-1} + \cdots + c_n S_{p-n} = 0$$

for $p > n$.

The standard proof of the first $n - 1$ formulas in this system uses techniques of calculus. The interested reader can find the proof in most any text on the theory of equations. The formulas for $p \geq n$ are immediate. They follow by summing the n equations

$$x_i^p + c_1 x_i^{p-1} + \cdots + c_n x_i^{p-n} = 0, \quad i = 1, 2, \ldots, n.$$

Here is a simple example of the use of Newton's formulas.

Example 2.24 *If*

$$\begin{aligned} x + y + z &= 1, \\ x^2 + y^2 + z^2 &= 2, \\ x^3 + y^3 + z^3 &= 3, \end{aligned}$$

determine the value of $x^4 + y^4 + z^4$.

Solution. Let

$$P(t) = (t - x)(t - y)(t - z) = t^3 + c_1 t^2 + c_2 t + c_3.$$

The relevant formulas are

$$\begin{aligned} S_1 + c_1 &= 0, \\ S_2 + c_1 S_1 + 2c_2 &= 0, \\ S_3 + c_1 S_2 + c_2 S_1 + 3c_3 &= 0, \\ S_4 + c_1 S_3 + c_2 S_2 + c_3 S_1 &= 0. \end{aligned}$$

Substituting $S_1 = 1$, $S_2, = 2$, $S_3 = 3$ and solving for c_1, c_2, c_3, using the first three equations, we find $c_1 = -1$, $c_2 = -1/2$, $c_3 = -1/6$. Thus x, y, z are the roots of

$$t^3 - t^2 - \frac{1}{2}t - \frac{1}{6} = 0.$$

We don't have to solve this equation in order to find S_4. Simply note that the fourth equation now reads $S_4 - 3 - 1 - \frac{1}{6} = 0$. Thus

$$x^4 + y^4 + z^4 = \frac{25}{6}. \quad \square$$

Looking back at Newton's formulas, we note that all power sums of the roots of the equation

$$t^n + c_1 t^{n-1} + \cdots + c_n = 0$$

are real if all of the coefficients $c_1, c_2, \ldots, c_n$ are real. This trivial observation is used to advantage in the final example of this section.

Example 2.25 (1980 USAMO) *Let*

$$G_n = a^n \sin(nA) + b^n \sin(nB) + c^n \sin(nC),$$

where a, b, c, A, B, C are real numbers and $A + B + C$ is a multiple of π. Prove that if $G_1 = G_2 = 0$, then $G_n = 0$ for every natural number n.

Solution. Let $z_1 = a(\cos A + i \sin A)$, $z_2 = b(\cos B + i \sin B)$, and $z_3 = c(\cos C + i \sin C)$. Using De Moivre's formula, we see that since $G_1 = G_2 = 0$ and $A + B + C$ is a multiple of π,

$$\text{Im}(z_1 + z_2 + z_3) = \text{Im}(z_1^2 + z_2^2 + z_3^2) = \text{Im}(z_1 z_2 z_3) = 0.$$

Thus σ_1 and σ_3 are real. Since $S_1 = z_1 + z_2 + z_3$ and $S_2 = z_1^2 + z_2^2 + z_3^2$ are real, so is

$$\sigma_2 = z_1 z_2 + z_2 z_3 + z_3 z_1 = \frac{S_1^2 - S_2}{2}.$$

Thus $\sigma_1, \sigma_2, \sigma_3$ are real and

$$P(z) = (z - z_1)(z - z_2)(z - z_3)$$

is a polynomial with real coefficients. Hence

$$S_n = z_1^n + z_2^n + z_3^n$$

is real for all $n \geq 0$. This gives us the desired conclusion since $G_n = \text{Im}\, S_n$. □

Exercises for Section 2.2

1. Solve
   (a) $x^4 - 2x^3 + 3x^2 - 2x - 3 = 0$,
   (b) $x^4 - 5x^2 - 6x - 5 = 0$,
   (c) $(x^2 - 4x)^2 + (x - 2)^2 = 10$.
2. Solve
   (a) $(x^2 - 3x)(x - 1)(x - 2) = 3$,
   (b) $8x^3 + 12x - 7 = 0$.
3. The equation $x^5 - 209x + 56 = 0$ has two roots whose product is 1. Find them.

4. One of the solutions of the equation $x^4 + ax^3 + x^2 + bx - 2 = 0$ is equal to $1 - \sqrt{2}$. Find the other solutions if both a and b are rational numbers.

5. Find all values of the parameter a such that all roots of the equation

$$x^6 + 3x^5 + (6-a)x^4 + (7-2a)x^3 + (6-a)x^2 + 3x + 1 = 0$$

are real. [1985 Bulgariam Olympiad]

6. Solve $x^5 - 3x^4 + x^3 + x^2 - 3x + 1 = 0$.

7. Solve $x^4 + 10x^2 + 12x + 40 = 0$.

8. If x, y, z satisfy

$$\begin{aligned} x + y + z &= 3, \\ x^2 + y^2 + z^2 &= 5, \\ x^3 + y^3 + z^3 &= 12, \end{aligned}$$

determine $x^4 + y^4 + z^4$.

9. If $x^2 + y^2 = 9$ and $x^3 + y^3 = 27$, determine all possible values (real or complex) of $x^4 + y^4$.

10. Let a, b, c be real numbers such that $a + b - c = 0$. Show that

$$\frac{a^5 + b^5 + c^5}{5} = \left(\frac{a^2 + b^2 + c^2}{2}\right)\left(\frac{a^3 + b^3 + c^3}{3}\right).$$

11. Find a cubic equation whose roots are the cubes of those of $x^3 + ax^2 + bx + c = 0$.

12. If the equation $x^3 + ax^2 + bx + c = 0$ has three real roots that are the lengths of the sides of a triangle, find a formula (in terms of a, b, c) for the area of the triangle. (The area of a triangle with sides x_1, x_2, x_3 is $\sqrt{s(s - x_1)(s - x_2)(s - x_3)}$ where $s = (x_1 + x_2 + x_3)/2$.)

13. Show that one of the roots of the equation $x^3 + ax^2 + bx + c = 0$ is the average of the other two if and only if $2a^3 - 9ab + 27c = 0$.

14. A student awoke at the end of an algebra class just in time to hear the teacher say, "...and I will give you a hint that the roots form an arithmetic progression." Looking at the board, he discovered a fifth degree equation to be solved for homework, which he hastily tried to copy down. He succeeded in getting only

$$x^5 - 5x^4 - 35x^3 + \cdots$$

before the teacher erased the blackboard. He was able to find all of the roots anyway. What are they?

15. Determine the power sums $S_1, S_2, \ldots, S_n$ for the roots of the equation

$$x^n + \frac{x^{n-1}}{1} + \frac{x^{n-2}}{2!} + \cdots + \frac{1}{n!} = 0.$$

2.3 Algebraic Equations and Inequalities

2.3.1 Solving Algebraic Equations

In his classic book *How to Solve It*, George Pólya gives what he calls a *Short Dictionary of Heuristic*. This is an alphabetical list of important strategies of mathematical problem solving. Polya's list includes the following:

Analogy
Cases
Contradiction
Extreme Conditions
Generalization
Patterns
Specialization
Symmetry
Variations
Working Backwards

One of the most basic strategies is that of reducing a new problem to one that we already know how to solve. Uses of this idea occur throughout mathematics. The particular example that concerns us here is the reduction of a non-polynomial algebraic equation to the solution of one or more polynomial equations.

The problem of solving a **rational equation** is just one small step away from that of solving a polynomial equation. The problem in question is to solve $P(x)/Q(x) = 0$, where P and Q are polynomials. The solution set is simply $S = \{x|P(x) = 0,\ Q(x) \neq 0\}$.

Example 2.26 *Find all solutions of*

$$x^2 + \frac{4x^2}{(x-2)^2} = 12.$$

Solution. This equation is equivalent to

$$x^2(x-2)^2 + 4x^2 = 12(x-2)^2, \qquad x \neq 2.$$

Since is it clear that $x = 2$ is not one of its roots, we simply have to solve the above fourth degree equation. Note that the equation can be written as

$$(x^2)^2 - 4x^2(x-2) - 12(x-2)^2 = 0,$$

and this factors by inspection to yield

$$\left(x^2 + 2(x-2)\right)\left(x^2 - 6(x-2)\right) = 0.$$

The desired solution set is thus $S = \{-1 \pm \sqrt{5},\ 3 \pm i\sqrt{3}\}$. □

Example 2.27 *Solve the rational equation*

$$\frac{2x(a+2)}{3(a-2x)} + \frac{1}{a} = \frac{2a+3}{a(a-2x)}.$$

Solution. The given equation is equivalent to

$$2a(a+2)x + 3(a-2x) - 3(2a+3) = 0, \qquad a \neq 0,\ a - 2x \neq 0.$$

Factoring, we obtain

$$(a+3)(2(a-1)x - 3) = 0, \qquad a \neq 0,\ a - 2x \neq 0.$$

If $a = 1$, there is no solution. Suppose that $a \neq 0, 1, -3$. Then

$$x = \frac{3}{2(a-1)}$$

is the unique solution *except* when this value of x leads to $a - 2x = 0$. The special values of a for which this is the case are the roots of $a^2 - a - 3 = 0$, namely

$$a = \frac{1 \pm \sqrt{13}}{2}.$$

Thus there are four values of a for which the solution set is empty: $a = 0,\ 1,\ (1 \pm \sqrt{13})/2$. There is one other special value of a, namely

$a = -3$. For $a = -3$, every $x \neq -3/2$ is a solution. In summary, the solution set $S(a)$ of the given rational equation is

$$S(a) = \begin{cases} \emptyset & \text{if } a = 0,\ 1,\ \frac{1\pm\sqrt{13}}{2}, \\ \{x|x \neq -\frac{3}{2}\} & \text{if } a = -3, \\ \left\{\frac{3}{2(a-1)}\right\} & \text{otherwise.} \end{cases} \qquad \square$$

Equations with radicals provide some new complications, but our basic strategy is still in effect. Given an equation of the form $f(x) = 0$ where f is an algebraic function involving radicals, we try to reduce the problem to one or more polynomial equations. We shall confine our attention to real solutions. In order to go from an algebraic equation involving radicals to a polynomial equation, we use two basic techniques:

- Raise both sides of the equation to the same power.
- Substitute for expressions involving radicals.

The first method has the disadvantage that the new equation might be satisfied by numbers that are *not* roots of the original. However, this is often the only available approach. If it is applicable, the second method may offer a more attractive solution. Sometimes it is helpful to combine the two methods.

Example 2.28 *Solve* $2(x-3) = \sqrt{x^2 - 2x - 3}$.

Solution. This is completely straightforward; just square both sides and solve. An x satisfying the given equation also satisfies the polynomial equation

$$4(x-3)^2 - (x-3)(x+1) = 0.$$

Factoring, we obtain

$$(x-3)(3x-13) = 0.$$

Thus the possible solutions are $x = 3$ and $x = 13/3$. (In squaring both sides, we might have introduced extra roots.) It is easy to check that both of these solutions are legitimate. Thus the solution set is $S = \{3,\ 13/3\}$. $\square$

Example 2.29 *Find all real solutions of $x = a + \sqrt{a + \sqrt{x}}$, where a is a real parameter.*

Solution. Suppose that x satisfies the given equation, and set $u = \sqrt{x}$. Then $x \geq 0$, $u \geq 0$, $u^2 \geq a$ and

$$(u^2 - a)^2 = a + u.$$

Thus u satisfies the fourth degree equation

$$u^4 - 2au^2 - u + a(a-1) = 0.$$

Good fortune is with us; we can factor this equation by inspection to get

$$\left(u^2 - u - a\right)\left(u^2 + u - (a-1)\right) = 0.$$

Since $u^2 \geq a$ and $u \geq 0$, the second factor does not vanish. Hence $u^2 - u - a = 0$, so x satisfies $x = a + \sqrt{x}$. Conversely, an x that satisfies $x = a + \sqrt{x}$ also satisfies the original equation. But x satisfies $x = a + \sqrt{x}$ if and only if $u = \sqrt{x}$ is a nonnegative root of $u^2 - u - a = 0$. Thus the solution set of the given equation is

$$S(a) = \begin{cases} \varnothing & \text{if } a < -\frac{1}{4}, \\ \left\{\left(\dfrac{1 \pm \sqrt{1+4a}}{2}\right)^2\right\} & \text{if } -\frac{1}{4} \leq a \leq 0, \\ \left\{\left(\dfrac{1 + \sqrt{1+4a}}{2}\right)^2\right\} & \text{if } a > 0. \end{cases} \qquad \square$$

The next example illustrates the substitution method.

Example 2.30 *Find all real solutions of $\sqrt[4]{x-1} + \sqrt[4]{5-x} = 2$.*

Solution. Let $s = \sqrt[4]{x-1}$ and $t = \sqrt[4]{5-x}$. Then s and t are nonnegative real numbers satisfying

$$\begin{aligned} s + t &= 2, \\ s^4 + t^4 &= 4. \end{aligned}$$

To find s and t, set $s = 1 + z$, $t = 1 - z$, and substitute into the second equation to get

$$z^4 + 6z^2 - 1 = 0.$$

The real solutions of this equation are

$$z = \pm\sqrt{\sqrt{10} - 3}.$$

Using the fact that $z^4 + 6z^2 = 1$, we find that x is related to z by $x = (1 + z)^4 + 1 = 3 + 4z(1 + z^2)$. Thus the solution set is

$$S = \left\{3 \pm 4(\sqrt{10} - 2)\sqrt{\sqrt{10} - 3}\right\}. \quad \square$$

The absolute value of x is defined by

$$|x| = \sqrt{x^2} = \begin{cases} x & \text{if } x \geq 0, \\ -x & \text{if } x \leq 0. \end{cases}$$

Equations involving absolute value usually require some case analysis, but this presents no serious difficulty. We are satisfied if each case reduces to a solvable equation.

Example 2.31 *Solve* $|x^2 - 3x| = x - 1$.

Solution. Using the definition of absolute value, we have

$$|x^2 - 3x| = \begin{cases} x^2 - 3x & \text{if } x \leq 0 \text{ or } x \geq 3, \\ 3x - x^2 & \text{if } 0 \leq x \leq 3. \end{cases}$$

From $|x^2 - 3x| = x - 1$ we obtain two quadratic equations: $x^2 - 4x + 1 = 0$ and $x^2 - 2x - 1 = 0$. Since only solutions with $x \geq 1$ can possibly satisfy $|x^2 - 3x| = x - 1$, it is easy to sort out the contenders and thus find the solution set $S = \{2 + \sqrt{3}, 1 + \sqrt{2}\}$. $\square$

Example 2.32 (1959 IMO) *For what real values of x is*

$$\sqrt{x + \sqrt{2x - 1}} + \sqrt{x - \sqrt{2x - 1}} = A,$$

given (a) $A = \sqrt{2}$, (b) $A = 1$, (c) $A = 2$, where only nonnegative real numbers are admitted for square roots?

Solution. The domain of the function on the left is $\{x | x \geq 1/2\}$. To see this, note that $x \geq 1/2$ is necessary because of the occurrence of $\sqrt{2x - 1}$. But $x - \sqrt{2x - 1} \geq 0$ for all $x \geq 1/2$, so the domain of the left hand side is $\{x | x \geq 1/2\}$. Squaring both sides of the equation, we obtain

$$2x + 2\sqrt{x^2 - (2x - 1)} = A^2,$$

which can be written as

$$x + |x - 1| = A^2/2.$$

(Since A is positive and it is understood that $x \geq 1/2$, no extraneous roots are introduced by squaring both sides.) Now

$$x + |x - 1| = \begin{cases} 2x - 1 & \text{if } x \geq 1, \\ 1 & \text{if } x \leq 1, \end{cases}$$

and we are ready to look at cases. For each value of A, there are two alternatives: (i) $2x - 1 = A^2/2$ and $x \geq 1$, (ii) $1 = A^2/2$ and $1/2 \leq x \leq 1$. Now it is easy to determine the solution set S in each case:

$$S = \begin{cases} \{x \mid 1/2 \leq x \leq 1\} & \text{if } A = \sqrt{2}, \\ \varnothing & \text{if } A = 1, \\ \{3/2\} & \text{if } A = 2. \end{cases} \qquad \square$$

2.3.2 Conditional Inequalities in One Variable

In this section, we describe some techniques for solving conditional inequalities in one variable. An inequality of the form $f(x) > 0$ where f is a polynomial or other algebraic function can be dealt with using basic facts about **continuity**. A theorem of Bolzano states that if $f(x)$ is continuous for $a \leq x \leq b$ and $f(a)$ is opposite in sign from $f(b)$ then there is a point c between a and b such that $f(c) = 0$. It follows that continuous real-valued functions have constant sign between consecutive real zeros. The assumption of continuity is not as restrictive as it might appear. Polynomials are everywhere continuous and rational functions are continous at all points where they are defined. (Specifically, the rational function $P(x)/Q(x)$ is continuous at every point where $Q(x) \neq 0$.)

Example 2.33 *Solve* $x^3 - 3x + 1 > 0$.

Solution. In the last section, we found that the roots of $x^3 - 3x + 1 = 0$ are $x_1 = 2\cos(8\pi/9)$, $x_2 = 2\cos(4\pi/9)$, and $x_3 = 2\cos(2\pi/9)$.

From elementary facts about the cosine function, these numbers are ordered as follows: $-2 < x_1 < 0 < x_2 < 1 < x_3 < 2$. Now it is a simple matter to test each interval to find the sign of $x^3 - 3x + 1$. Thus we find that the inequality is satisfied if $2\cos(8\pi/9) < x < 2\cos(4\pi/9)$ or $x > 2\cos(2\pi/9)$. □

Sometimes we are fortunate enough to find a more direct approach to the solution of an inequality.

Example 2.34 (1960 IMO) *For what values of the variable x does the following inequality hold:*

$$\frac{4x^2}{(1-\sqrt{1+2x})^2} < 2x + 9?$$

Solution. The left-hand side is defined if $x \geq -1/2$ and $x \neq 0$. If these conditions are satisfied, the left-hand side can be written in a far simpler form. Just multiply numerator and denominator by $(1 + \sqrt{1+2x})^2$ and simplify. We thus get the following equivalent problem:

$$2\sqrt{2x+1} < 7, \quad x \geq -1/2,\ x \neq 0.$$

This inequality is satisfied if $x \neq 0$ and $0 \leq 2x + 1 < 49/4$. Thus the solution set is $S = \{x \mid -1/2 \leq x \leq 45/8, \quad x \neq 0\}$. □

Exercises for Section 2.3

For each equation or inequality that appears without further comment, find all real solutions.

1. Find all solutions of

$$\frac{(x+a)(x+b)}{(x-a)(x-b)} + \frac{(x+b)(x+c)}{(x-b)(x-c)} + \frac{(x+c)(x+a)}{(x-c)(x-a)} = 3,$$

where a, b, c are real parameters.

2. $\sqrt{x+1} - 1 = \sqrt{\frac{x-1}{x}}$.

3. $x^3 - 4x^2 + 3x + 2(x-1)\sqrt{x} = 0$.

4. $x + 3\sqrt[3]{x} = 1$.

5. Find all real solutions of $\sqrt{x+1} - \sqrt{x} = a$, where a is a real parameter.

6. $\sqrt{4x + \sqrt{7x+2}} - 2\sqrt{x} = 1$.

7. $\sqrt[3]{x} + \sqrt[3]{x - 16} = \sqrt[3]{x - 8}$.
8. $\sqrt[4]{x - 2} + \sqrt[4]{3 - x} = 1$.
9. $|2x^2 - x - 1| = x$.
10. $\sqrt{x^2 - 2|x| + 1} = x^2 - 4|x - 1| + 1$.
11. Find all real solutions of $|2x - a| = |x + 1| - x + 4$, where a is a real parameter.
12. Find all real numbers c such that the quadratic polynomial $P(x) = x^2 + cx + c(c + 6)$ satisfies $P(x) < 0$ for $1 < x < 2$.
13. $\sqrt{\sqrt{x} + 8} - \sqrt{x} \geq 2$.
14. $|x - 2||x - 4| \leq 8x^2 - x + 3$.
15. Solve the inequality $\sqrt{x^2 + a^2} > x + a - 1$, where a is a real parameter.

2.4 The Classical Inequalities

Sometimes a very simple observation in mathematics leads to an explosion of new ideas and useful results. An observation that provides the genesis of many important inequalities is simply that the square of any real number is nonnegative. If a and b are nonnegative numbers and we apply this observation to $\sqrt{a} - \sqrt{b}$, we find

$$(\sqrt{a} - \sqrt{b})^2 = a + b - 2\sqrt{ab} \geq 0,$$

or

$$\frac{a + b}{2} \geq \sqrt{ab}. \tag{2.5}$$

Starting with

$$(x - y)^2 + (y - z)^2 + (z - x)^2 \geq 0,$$

we find

$$x^2 + y^2 + z^2 - (xy + yz + zx) \geq 0.$$

Thus

$$(x + y + z)\left[(x^2 + y^2 + z^2) - (xy + yz + zx)\right] \geq 0$$

for all $x, y, z \geq 0$. Multiplying out the left-hand side, we discover that this inequality is just

$$x^3 + y^3 + z^3 \geq 3xyz, \qquad x, y, z \geq 0.$$

Letting $x = \sqrt[3]{a}$, $y = \sqrt[3]{b}$, $z = \sqrt[3]{z}$, we obtain

$$\frac{a+b+c}{3} \geq \sqrt[3]{abc}. \tag{2.6}$$

Inequalties (2.5) and (2.6) are two special cases of the famous **arithmetic mean - geometric mean inequality**: for any n nonnegative real numbers $a_1, a_2, \ldots, a_n$,

$$\frac{a_1 + a_2 + \cdots + a_n}{n} \geq \sqrt[n]{a_1 a_2 \cdots a_n},$$

with equality only when all of the numbers are equal. We shall henceforth refer to this result as the **AM-GM inequality**. The left side is the arithmetic mean (A) and the right-side is the geometric mean (G). These are two kinds of "averages" of the numbers $a_1, a_2, \ldots, a_n$. There are others; if all a_i are positive, the **harmonic mean** is defined by

$$H = \frac{n}{\frac{1}{a_1} + \frac{1}{a_2} + \cdots + \frac{1}{a_n}},$$

in other words, H is the reciprocal of the arithmetic mean of the reciprocals. Using AM-GM, it is not hard to show that the harmonic mean is always less than or equal to the geometric mean, so $H \leq G \leq A$. This and much more is contained in the following general result.

Theorem 2.8 (Power Mean Inequality) *Let $a_1, a_2, \ldots, a_n$ be positive real numbers, and let α be real. Let*

$$M_\alpha(a_1, \ldots, a_n) = \left(\frac{a_1^\alpha + a_2^\alpha + \cdots + a_n^\alpha}{n}\right)^{1/\alpha} \qquad (\alpha \neq 0)$$

and

$$M_0(a_1, a_2, \ldots, a_n) = \sqrt[n]{a_1 a_2 \cdots a_n}.$$

Then M_α is an increasing function of α unless $a_1 = a_2 = \cdots = a_n$ (in which case M_α is constant).

By defining M_0 as the geometric mean, we ensure that M_α is a continuous function of α. Special power means are $M_{-1} = H$ (harmonic mean), $M_0 = G$ (geometric mean), and $M_1 = A$ (arithmetic mean). AM-GM and other cases of the power mean inequality are extremely useful. Another mainstay is **Cauchy's inequality**.

Note. In different books, you may see this called the Cauchy-Schwarz or even the Cauchy-Schwarz-Buniakowsky inequality. However, Schwarz and Buniakowsky produced *generalizations* of Cauchy's inequality. In the form in which we shall state it, the inequality is due to Cauchy.

Theorem 2.9 (Cauchy's Inequality) *For arbitrary real numbers* $a_1, a_2,$
$\ldots, a_n$ *and* $b_1, b_2, \ldots, b_n$,

$$(a_1b_1 + \cdots + a_nb_n)^2 \le (a_1^2 + \cdots + a_n^2)(b_1^2 + \cdots + b_n^2).$$

Furthermore, equality holds only when there are numbers λ, μ, *not both zero, such that* $\lambda a_i = \mu b_i$ *for all* i. *(We say in this case that the n-tuples* $(a_1, a_2, \ldots, a_n)$ *and* $(b_1, b_2, \ldots, b_n)$ *are* **proportional**.*)*

An application of Cauchy's inequality gives the following basic result, known as the **triangle inequality**. To set the stage, let $\mathbb{R}^n$ denote the set of all n-tuples $x = (x_1, x_2, \ldots, x_n)$ where $x_1, x_2, \ldots, x_n$ are real numbers. The elements of $\mathbb{R}^n$ are called **vectors**. In $\mathbb{R}^n$ vectors are added by the rule

$$(x_1, x_2, \ldots, x_n) + (y_1, y_2, \ldots, y_n) = (x_1 + y_1, x_2 + y_2, \ldots, x_n + y_n).$$

Assign to each vector x a **length**, or **norm**, $\|x\|$ given by

$$\|x\| = (x_1^2 + x_2^2 + \cdots + x_n^2)^{1/2}.$$

Theorem 2.10 (Triangle Inequality) *For any two vectors* x, y *in* $\mathbb{R}^n$,

$$\|x + y\| \le \|x\| + \|y\|.$$

Equality holds if and only if $(x_1, x_2, \ldots, x_n)$ *and* $(y_1, y_2, \ldots, y_n)$ *are proportional.*

To prove this, just write out the square of the left-hand side and apply Cauchy's inequality:

$$\|x + y\|^2 = (x_1 + y_1)^2 + (x_2 + y_2)^2 + \cdots + (x_n + y_n)^2$$

$$\begin{aligned}
&= (x_1^2 + \cdots + x_n^2) + (y_1^2 + \cdots + y_n^2) \\
&\quad + 2(x_1y_1 + \cdots + x_ny_n) \\
&\leq (x_1^2 + \cdots + x_n^2) + (y_1^2 + \cdots + y_n^2) \\
&\quad + 2\sqrt{x_1^2 + \cdots + x_n^2}\sqrt{y_1^2 + \cdots + y_n^2} \\
&= (\|x\| + \|y\|)^2.
\end{aligned}$$

This is easily generalized by mathematical induction to yield

$$\|u_1 + u_2 + \cdots + u_k\| \leq \|u_1\| + \|u_2\| + \cdots + \|u_k\|$$

for arbitrary vectors $u_1, u_2, \ldots, u_k$ in $\mathbb{R}^n$. Note that for the case of real numbers ($n = 1$), the norm is just the absolute value.

Many interesting results can be proved easily by using AM-GM or Cauchy's inequality.

Example 2.35 *Show that if a, b, c are nonnegative real numbers, then*

$$(a + b)(b + c)(c + a) \geq 8abc.$$

Solution. We know that $a + b \geq 2\sqrt{ab}$, $b + c \geq 2\sqrt{bc}$, and $c + a \geq 2\sqrt{ca}$. Thus

$$(a + b)(b + c)(c + a) \geq 8\sqrt{ab}\sqrt{bc}\sqrt{ca} = 8abc.$$

Equality holds if and only if $a = b = c$. □

Example 2.36 (1975 Putnam Competition) *Let*

$$H_n = 1 + \frac{1}{2} + \cdots + \frac{1}{n}.$$

Show that

$$n(n + 1)^{1/n} < n + H_n$$

for every $n > 1$.

Solution. This is another job for the AM-GM inequality. Rewriting the inequality to be proved as

$$\frac{n + H_n}{n} > (n + 1)^{1/n},$$

we get a good clue. Let $a_k = 1 + \frac{1}{k}$ for $k = 1, 2, \ldots, n$ and apply the AM-GM inequality to the numbers $a_1, a_2, \ldots, a_n$. We have

$$A = \frac{(1 + 1) + (1 + \frac{1}{2}) + \cdots + (1 + \frac{1}{n})}{n} = \frac{n + H_n}{n}$$

and

$$G = \left\{(2)\left(\frac{3}{2}\right)\cdots\left(\frac{n+1}{n}\right)\right\}^{1/n} = (n+1)^{1/n}.$$

Since $n > 1$ the terms $a_1, \ldots, a_n$ are unequal, and thus the inequality is strict. □

Example 2.37 *Prove that for positive real numbers a, b, c, d,*

$$\frac{a}{b+c} + \frac{b}{c+d} + \frac{c}{d+a} + \frac{d}{a+b} \geq 2.$$

Solution. Let $S = a + b + c + d$. By AM-GM, we have

$$\frac{a}{b+c} + \frac{c}{d+a} = \frac{a(d+a) + c(b+c)}{(b+c)(d+a)} \geq \frac{a(d+a) + c(b+c)}{(S/2)^2}$$

and

$$\frac{b}{c+d} + \frac{d}{a+b} = \frac{b(a+b) + d(c+d)}{(a+b)(c+d)} \geq \frac{b(a+b) + d(c+d)}{(S/2)^2}.$$

Thus

$$\frac{a}{b+c} + \frac{b}{c+d} + \frac{c}{d+a} + \frac{d}{a+b} \geq \frac{4P}{S^2},$$

where

$$P = ad + a^2 + bc + c^2 + ab + b^2 + cd + d^2.$$

Notice that in the expression defining P, the square terms a^2, b^2, c^2, and d^2 each occur and that the "missing" cross terms are ac and bc. This motivates the observation that

$$P = \frac{1}{2}\left[S^2 + (a-c)^2 + (b-d)^2\right].$$

It follows that $4P \geq 2S^2$, so

$$\frac{a}{b+c} + \frac{b}{c+d} + \frac{c}{d+a} + \frac{d}{a+b} \geq 2,$$

as claimed. A brief check of the steps shows that equality holds if and only if $a = c$ and $b = d$. □

Often a specific inequality can be deduced in various ways using one or more of the classical inequalities. A case in point is the inequality

$$(a_1 + a_2 + \cdots + a_n)\left(\frac{1}{a_1} + \frac{1}{a_2} + \cdots + \frac{1}{a_n}\right) \geq n^2, \qquad (2.7)$$

which is valid for positive $a_1, \ldots, a_n$. On the one hand, this inequality follows that between the arithmetic and harmonic means, $M_{-1} \leq M_1$. On the other hand, the same inequality comes from applying Cauchy's inequality to the sequences $(\sqrt{a_1}, \sqrt{a_2}, \ldots, \sqrt{a_n})$ and $(\frac{1}{\sqrt{a_1}}, \frac{1}{\sqrt{a_2}}, \ldots, \frac{1}{\sqrt{a_n}})$.

Note that the AM-GM inequality can be written as

$$a_1 a_2 \cdots a_n \leq \left(\frac{a_1 + a_2 + \cdots + a_n}{n}\right)^n.$$

In other words, the product of n nonnegative numbers does not exceed the nth power of their arithmetic mean. This statement has a useful interpretation as a constrained maximum problem. If the sum of n nonnegative numbers is S, then the maximum possible value for the product of these numbers is $(S/n)^n$. One of the standard applications of calculus has to do with finding the maximum (or minimum) value of a given function. In cases where the problem comes down to finding the maximum value of a product of nonnegative expressions whose sum is fixed, calculus is often unnecessary. The desired result may be obtained through a simple application of the AM-GM inequality.

Example 2.38 *Find the maximum possible value of* $x(1 - x^3)$ *for* $0 \leq x \leq 1$.

Solution. The idea is to write the expression to be maximized as a product of expressions whose sum is a constant. Let $y = x(1 - x^3)$. Then

$$y^3 = x^3(1 - x^3)^3.$$

Multiplying by 3, we obtain

$$3y^3 = (3x^3)(1 - x^3)(1 - x^3)(1 - x^3),$$

in which the right-hand side is the product of four expressions whose sum is constant. Since $0 \leq x \leq 1$, each factor is nonnegative. The

AM-GM inequality now yields

$$3y^3 \leq \left(\frac{3}{4}\right)^4.$$

Thus

$$y \leq \frac{3}{16}\sqrt[3]{16},$$

and equality holds only in case $3x^3 = 1 - x^3$. Thus the maximum value of $x(1 - x^3)$ for $0 \leq x \leq 1$ occurs at $x = \sqrt[3]{1/4}$. □

Many inequality problems can be solved very elegantly by using facts concerning **convex functions**. Suppose that f is a continuous real-valued function that is defined on an interval, and that f has the following property: for any two points x, y in the interval,

$$f\left(\frac{x+y}{2}\right) \leq \frac{f(x) + f(y)}{2}.$$

Then f is **convex**. Geometrically, this means that the mid-point of every chord lies on or above the corresponding point on the graph of the function. *Note.* By requiring for all λ between 0 and 1 that $f\{\lambda x + (1-\lambda)y\} \leq \lambda f(x) + (1-\lambda)f(y)$, one can omit the requirement that f is continuous; it is automatically satisfied. However, the above definition suits our purposes. As readers who have studied calculus may already know, f is convex on an interval in which f'' is nonnegative.

Example 2.39 *Show that if*

$$f(x) = \frac{C}{x+r},$$

where $C > 0$ and r are constants, then f is convex for $x > -r$.

Solution. Two applications of the AM-GM inequality give

$$\frac{1}{2}\left[\frac{C}{x+r} + \frac{C}{y+r}\right] \geq \frac{C}{\sqrt{(x+r)(y+r)}} \geq \frac{C}{\frac{x+y}{2} + r},$$

and this shows that f is convex. □

Convex functions have some very useful properties.

Property 2.1 *If $w_1, w_2, \ldots, w_n$ are positive numbers satisfying $w_1 + w_2 + \cdots + w_n = 1$, and $x_1, x_2, \ldots, x_n$ are any n points in an interval* *where f is continuous and convex, then*

$$f(w_1x_1 + \cdots + w_nx_n) \le w_1f(x_1) + \cdots + w_nf(x_n).$$

Property 2.2 *If f is continuous and convex on $[a, b]$, then the maximum value of f is taken at an end-point, in other words if $M = \max\{f(a), f(b)\}$ then*

$$f(x) \le M$$

for all x in $[a, b]$.

Property 2.1 is called **Jensen's inequality**. A simple geometrical feature of convex functions lies at the heart of Jensen's inequality. The feature in question is that at every point $(a, f(a))$ on the graph of a convex function, there is at least one **supporting line**, that is to say a line of the form $y = f(a) + m(x - a)$ such that

$$f(x) \ge f(a) + m(x - a)$$

for all x. (More precisely, if f is convex in the interval I then the above inequality holds for every x in I.) Set $a = w_1x_1 + \cdots + w_nx_n$. Then a belongs to the interval of convexity. Using the support line inequality and the fact that $\sum_{i=1}^{n} w_i = 1$, we have Jensen's inequality:

$$\sum_{i=1}^{n} w_if(x_i) \ge \sum_{i=1}^{n} w_i\{f(a) + m(x_i - a)\} = f(a) = f\left(\sum_{i=1}^{n} w_ix_i\right).$$

To see the power of Jensen's inequality, note that by taking the logarithm of both sides of

$$\frac{x + y}{2} \ge \sqrt{xy}, \qquad x, y > 0,$$

we establish that $f(x) = -\log(x)$ is convex for $x > 0$. (It is a good idea to sketch the graph of $-\log(x)$ to gain an intuitive understanding of the last statement.) Setting $w_1 = w_2 = \cdots = w_n = \frac{1}{n}$ and using Property 2.1, we obtain

$$\log\left(\frac{x_1 + x_2 + \cdots + x_n}{n}\right) \ge \frac{\log x_1 + \log x_2 + \cdots + \log x_n}{n},$$

and so

$$\frac{x_1 + x_2 + \cdots + x_n}{n} \ge \sqrt[n]{x_1x_2\cdots x_n},$$

the AM-GM inequality. More generally, we have the following extension of AM-GM.

Theorem 2.11 (Weighted AM-GM Inequality) *If $x_1, x_2, \ldots, x_n$ are nonnegative real numbers and $w_1, w_2, \ldots, w_n$ are positive numbers satisfying $w_1 + w_2 + \cdots + w_n = 1$, then*

$$\prod_{i=1}^{n} x_i^{w_i} \leq \sum_{i=1}^{n} w_i x_i.$$

Equality holds if and only if $x_1 = x_2 = \cdots = x_n$.

Example 2.40 (1992 Chinese Mathematical Olympiad) *For every positive integer $n \geq 2$ find the smallest positive number $\lambda = \lambda(n)$ such that if*

$$0 \leq a_1, a_2, \ldots, a_n \leq \frac{1}{2} \qquad \text{and} \qquad b_1, b_2, \ldots, b_n > 0$$

satisfy $a_1 + a_2 + \cdots + a_n = b_1 + b_2 + \cdots + b_n = 1$ then

$$b_1 b_2 \cdots b_n \leq \lambda(a_1 b_1 + a_2 b_2 + \cdots + a_n b_n).$$

Solution. Let

$$f(x) = \frac{b_1 b_2 \cdots b_n}{x}, \qquad x > 0.$$

Then f is convex on its domain $(0, \infty)$ (see Example 2.39), and since the a_i are nonnegative and satisfy $\sum_{i=1}^{n} a_i = 1$, Jensen's inequality shows that

$$\frac{b_1 b_2 \cdots b_n}{a_1 b_1 + \cdots + a_n b_n} = f\left(\sum_{i=1}^{n} a_i b_i\right) \leq \sum_{i=1}^{n} a_i f(b_i).$$

We may assume that the b_i have been indexed so that $b_1 \leq b_2 \leq \cdots \leq b_n$. Then $f(b_1) \geq f(b_2) \geq \cdots \geq f(b_n)$ and

$$\begin{aligned}\frac{b_1 b_2 \cdots b_n}{a_1 b_1 + \cdots + a_n b_n} &\leq a_1 f(b_1) + a_2 f(b_2) + \cdots + a_n f(b_n)\\ &\leq a_1 f(b_1) + (a_2 + \cdots + a_n) f(b_2)\\ &= a_1 f(b_1) + (1 - a_1) f(b_2)\\ &= f(b_2) + a_1(f(b_1) - f(b_2))\\ &\leq f(b_2) + \frac{1}{2}(f(b_1) - f(b_2))\end{aligned}$$

$$= \frac{1}{2}(b_1 + b_2)b_3b_4 \cdots b_n$$
$$\le \frac{1}{2}\left(\frac{1}{n-1}\right)^{n-1},$$

where the last step uses the AM-GM inequality. A brief check shows that equality holds if

$$a_1 = a_2 = 1/2, \quad a_3 = \cdots = a_n = 0$$

and

$$b_1 = b_2 = \frac{1}{2(n-1)}, \quad b_3 = \cdots = b_n = \frac{1}{n-1}.$$

Thus

$$\lambda(n) = \frac{1}{2}\left(\frac{1}{n-1}\right)^{n-1}$$

is the smallest λ such that

$$b_1b_2 \cdots b_n \le \lambda(a_1b_1 + a_2b_2 + \cdots + a_nb_n). \quad \square$$

Property 2.2 is used in the solution of the following problem.

Example 2.41 (1980 USAMO) *Show that if* $0 \le a, b, c \le 1$, *then*

$$\frac{a}{b+c+1} + \frac{b}{c+a+1} + \frac{c}{a+b+1} + (a-1)(b-1)(c-1) \le 1.$$

Solution. This is the kind of problem that may lead to grief if attacked by random manipulation. We must have an insightful approach if we are not to get lost in unending algebra. Denote the left side of the inequality by $f(a, b, c)$. The crucial observation is that f is continuous and also convex in each variable separately, in other words, if two of the three numbers a, b, c are fixed, then f is a convex function of the third quantity. Once this is established, the problem is solved since we only need to consider the case where a, b, c take extreme values to find the maximum value of f, and we find that for each of the eight points $\{(0,0,0), (1,0,0), \ldots, (1,1,1)\}$ the value of f is 1. Clearly f is continuous for $0 \le a, b, c \le 1$. To show that f is convex, we use the easily proved fact that any finite sum of convex functions is itself convex and note that of the four terms, two are linear in the chosen variable and the other two are of the form that was shown to be convex in Example 2.39. Thus f is convex in each variable, and now it follows that $f(a,b,c) \le 1$ for $0 \le a,b,c \le 1$. □

Theorem 2.12 *If a and b are nonnegative numbers and $p, q > 1$ satisfy $1/p + 1/q = 1$ then*

$$\frac{a^p}{p} + \frac{b^q}{q} \ge ab,$$

with equality if and only if $a^p = b^q$.

This is an immediate consequence of the weighted AM-GM inequality. Let $n = 2$ and set $x_1 = a^p$, $x_2 = b^q$ and $w_1 = 1/p$, $w_2 = 1/q$. Then

$$\sum_{i=1}^{n} w_i x_i \ge \prod_{i=1}^{n} x_i^{w_i}$$

yields

$$\frac{a^p}{p} + \frac{b^q}{q} \ge (a^p)^{1/p}(b^q)^{1/q} = ab.$$

Example 2.42 *Let n be an integer. Prove that if*

$$\frac{\sin^{2n+2} A}{\sin^{2n} B} + \frac{\cos^{2n+2} A}{\cos^{2n} B} = 1$$

holds for some $n \ne 0, -1$, then it holds for all $n \in \mathbb{Z}$.

Solution. Since the left-hand side is unchanged in value if we replace n by $-(n+1)$ and simultaneously interchange A and B, we may assume that $n \ge 1$. In this case, $p = n+1$ and $q = (n+1)/n$ are positive and satisfy $1/p + 1/q = 1$. Set

$$a = \frac{\sin^2 A}{(\sin^2 B)^{1/q}} \quad \text{and} \quad b = (\sin^2 B)^{1/q}.$$

Then Theorem 2.12 yields

$$\frac{1}{p}\frac{\sin^{2n+2} A}{\sin^{2n} B} + \frac{1}{q}\sin^2 B \ge \sin^2 A. \tag{2.8}$$

In exactly the same way,

$$\frac{1}{p}\frac{\cos^{2n+2} A}{\cos^{2n} B} + \frac{1}{q}\cos^2 B \ge \cos^2 A. \tag{2.9}$$

Since the sum of (2.8) and (2.9) yields

$$\frac{\sin^{2n+2} A}{\sin^{2n} B} + \frac{\cos^{2n+2} A}{\cos^{2n} B} \geq 1,$$

and we are given that

$$\frac{\sin^{2n+2} A}{\sin^{2n} B} + \frac{\cos^{2n+2} A}{\cos^{2n} B} = 1,$$

equality must hold in both (2.8) and (2.9). But the condition for equality ($a^p = b^q$) implies that $\sin^{2n+2} A = \sin^{2n+2} B$ and so $\sin^2 A = \sin^2 B$. Thus

$$\frac{\sin^{2k+2} A}{\sin^{2k} B} + \frac{\cos^{2k+2} A}{\cos^{2k} B} = \sin^2 A + \cos^2 A = 1$$

for every integer k. □

Theorem 2.12 can be used to prove the following classical inequality, due to Hölder.

Theorem 2.13 (Hölder's Inequality) *If $x_1, x_2, \ldots, x_n$ and $y_1, y_2, \ldots, y_n$ are nonnegative numbers and $p, q > 1$ satisfy $1/p + 1/q = 1$ then*

$$\sum_{i=1}^{n} x_i y_i \leq \left(\sum_{i=1}^{n} x_i^p\right)^{1/p} \left(\sum_{i=1}^{n} y_i^q\right)^{1/q}.$$

Equality holds if and only if $(x_1^p, x_2^p, \ldots, x_n^p)$ and $(y_1^q y_2^q, \ldots, y_n^q)$ are proportional.

Let $u = (\sum_{i=1}^n x_i^p)^{1/p}$ and $v = (\sum_{i=1}^n y_i^q)^{1/q}$. By Theorem 2.12,

$$\frac{x_i}{u}\frac{y_i}{v} \leq \frac{1}{p}\left(\frac{x_i}{u}\right)^p + \frac{1}{q}\left(\frac{y_i}{v}\right)^q, \qquad i = 1, 2, \ldots, n.$$

Summing on i and then multiplying by uv, we obtain Hölder's inequality.

Note that $p = q = 1/2$ in this result gives us Cauchy's inequality. Using Hölder's inequality, we can prove the following inequality of Minkowski generalizing the triangle inequality.

Theorem 2.14 (Minkowski's Inequality) *If $x_1, x_2, \ldots, x_n$ and $y_1, y_2, \ldots, y_n$ are nonnegative numbers and $p > 1$, then*

$$\left(\sum_{i=1}^{n} (x_i + y_i)^p\right)^{1/p} \leq \left(\sum_{i=1}^{n} x_i^p\right)^{1/p} + \left(\sum_{i=1}^{n} y_i^p\right)^{1/p}.$$

Equality holds if and only if $(x_1, x_2, \ldots, x_n)$ *and* $(y_1, y_2, \ldots, y_n)$ *are proportional*

Write

$$\sum_{i=1}^{n}(x_i + y_i)^p = \sum_{i=1}^{n} x_i(x_i + y_i)^{p-1} + \sum_{i=1}^{n} y_i(x_i + y_i)^{p-1},$$

and apply Hölder's inequality to both sums on the right. Note that $1/p + 1/q = 1$ yields $(p-1)q = p$.

Theorem 2.15 (Bernoulli's inequality) *If* $x > -1$ *and* n *is a positive integer, then*

$$(1 + x)^n \geq 1 + nx.$$

For real exponents, the following version holds: if $x > -1$ *and* $x \neq 0$ *then* $(1 + x)^r > 1 + rx$ *if* $r > 1$ *or* $r < 0$, *and* $(1 + x)^r < 1 + rx$ *if* $0 < r < 1$.

The 1991 USAMO provided an opportunity to use Bernoulli's inequality.

Example 2.43 (1991 USAMO) *Let*

$$a = (m^{m+1} + n^{n+1})/(m^m + n^n)$$

where m *and* n *are positive intgers. Prove that* $a^m + a^n \geq m^m + n^n$.

Solution. We apply Bernoulli's inequality to each of the expressions

$$\left(1 + \frac{a-m}{m}\right)^m \quad \text{and} \quad \left(1 + \frac{a-n}{n}\right)^n$$

to obtain

$$\begin{aligned} a^m + a^n &= m^m\left(1 + \frac{a-m}{m}\right)^m + n^n\left(1 + \frac{a-n}{n}\right)^n \\ &\geq m^m(1 + (a-m)) + n^n(1 + (a-n)) \\ &= m^m + n^n + a(m^m + n^n) - (m^{m+1} + n^{n+1}) \\ &= m^m + n^n. \quad \square \end{aligned}$$

As we have seen, some of the most important inequalities involve symmetric functions. Occasionally, the symmetry is somewhat hidden, and it is necessary to do some algebraic calculations to uncover it.

Example 2.44 (1992 Canadian Mathematical Olympiad) *For $x, y, z \geq 0$, establish the inequality*

$$x(x-z)^2 + y(y-z)^2 \geq (x-z)(y-z)(x+y-z),$$

and determine when equality holds.

Solution. Write the inequality to be proved as

$$x(x-z)^2 + y(y-z)^2 - (x-z)(y-z)(x+y-z) \geq 0.$$

Grouping terms, the left-hand side can be written

$$x(x-z)[(x-z)-(y-z)] + y(y-z)[(y-z)-(x-z)] + z(x-z)(y-z)$$

$$= x(x-y)(x-z) + y(y-x)(y-z) + z(z-x)(z-y),$$

which is manifestly symmetric in x, y, z. In view of symmetry, it suffices to prove the original inequality for $x \geq z \geq y \geq 0$. The inequality is obviously true in this case since

$$x(x-z)^2 + y(y-z)^2 \geq 0 \geq (x-z)(y-z)(x+y-z). \quad \square$$

Let $X = [x_{ij}]$ be an $m \times n$ matrix of nonnegative real numbers, and assume that no column of X consists entirely of zeros. By the weighted AM-GM inequality,

$$\prod_{j=1}^{n} \left(\frac{x_{ij}}{u_j}\right)^{w_j} \leq \sum_{j=1}^{n} w_j \left(\frac{x_{ij}}{u_j}\right), \quad i = 1, 2, \ldots, m,$$

where

$$u_j = x_{1j} + x_{2j} + \cdots + x_{mj}, \quad j = 1, 2, \ldots, n.$$

Summing on i we obtain the following generalization of Theorem 2.13.

Theorem 2.16 (Hölder) *Let $X = [x_{ij}]$ be an $m \times n$ matrix of nonnegative real numbers and let $w_1, w_2, \ldots, w_n$ be positive numbers satisfying $w_1 + w_2 + \cdots + w_n = 1$. Then*

$$\sum_{i=1}^{m} \prod_{j=1}^{n} x_{ij}^{w_j} \leq \prod_{j=1}^{n} \left(\sum_{i=1}^{m} x_{ij}\right)^{w_j},$$

with equality if and only if every two columns of X are proportional or one column consists entirely of zeros.

This result is used to prove a mixed arithmetic mean-geometric mean inequality in the following example.

Example 2.45 *Prove that for* $a, b, c > 0$,

$$\frac{a+\sqrt{ab}+\sqrt[3]{abc}}{3} \le \sqrt[3]{a\cdot\frac{a+b}{2}\cdot\frac{a+b+c}{3}}.$$

Solution. Let

$$X = \begin{bmatrix} a & a & a \\ a & \sqrt{ab} & b \\ a & b & c \end{bmatrix}$$

and set $w_1 = w_2 = w_3 = 1/3$. From Theorem 2.16, we have

$$\frac{1}{3}\sum_{i=1}^{3}\prod_{j=1}^{3} x_{ij}^{1/3} \le \frac{1}{3}\prod_{j=1}^{3}\left(\sum_{i=1}^{3} x_{ij}\right)^{1/3},$$

and this gives

$$\frac{a+\sqrt{ab}+\sqrt[3]{abc}}{3} \le \frac{1}{3}\sqrt[3]{3a(a+b+\sqrt{ab})(a+b+c)}.$$

Using $\sqrt{ab} \le (a+b)/2$, we have

$$3a(a+b+\sqrt{ab})(a+b+c) \le 27a\left(\frac{a+b}{2}\right)\left(\frac{a+b+c}{3}\right),$$

so

$$\frac{a+\sqrt{ab}+\sqrt[3]{abc}}{3} \le \sqrt[3]{a\cdot\frac{a+b}{2}\cdot\frac{a+b+c}{3}}.$$

Equality holds if and only if $a = b = c$. □

Note. This is a special case of the following mixed arithmetic mean-geometric mean inequality, proved by Kiran Kedlaya.

Theorem 2.17 (Kedlaya) *Let* $a_1, a_2, \ldots, a_n$ *be positive real numbers. The arithmetic mean of the numbers*

$$a_1,\ \sqrt{a_1a_2},\ \sqrt[3]{a_1a_2a_3},\ \ldots,\ \sqrt[n]{a_1a_2\cdots a_n}$$

does not exceed the geometric mean of the numbers

$$a_1,\ \frac{a_1+a_2}{2},\ \frac{a_1+a_2+a_3}{3},\ \ldots,\ \frac{a_1+a_2+\cdots+a_n}{n}.$$

There is equality if and only if $a_1 = a_2 = \cdots = a_n$.

The proof is obtained by the ingenious construction of a system of weights $\epsilon_k(i,j)$ such that by applying the generalized Hölder inequality in the case of the geometric means

$$x_{ij} = \prod_{k=1}^{n} a_k^{\epsilon_k(i,j)}, \quad 1 \le i,j \le n$$

and setting $w_1 = w_2 = \cdots = w_n = 1/n$, one obtains

$$\frac{1}{n}\sum_{i=1}^{n}\prod_{j=1}^{n} x_{ij}^{1/n} = \frac{a_1 + \sqrt{a_1a_2} + \cdots + \sqrt[n]{a_1a_2\cdots a_n}}{n}.$$

At the same time,

$$\frac{1}{n}\prod_{j=1}^{n}\left(\sum_{i=1}^{n} x_{ij}\right)^{1/n} \le \sqrt[n]{a_1\left(\frac{a_1+a_2}{2}\right)\cdots\left(\frac{a_1+a_2+\cdots+a_n}{n}\right)}$$

through an application of the weighted AM-GM inequality.

Theorem 2.18 (Rearrangement Inequality) *Let $a_1, a_2, \ldots, a_n$ and $b_1, b_2, \ldots, b_n$ be two sequences of real numbers, and suppose $a_1 \le a_2 \le \cdots \le a_n$. For each permutation π of $\{1, 2, \ldots, n\}$ let*

$$\Sigma(\pi) = \sum_{k=1}^{n} a_k b_{\pi(k)}.$$

Then Σ is largest when $b_{\pi(1)} \le b_{\pi(2)} \le \cdots \le b_{\pi(n)}$ and smallest when $b_{\pi(1)} \ge b_{\pi(2)} \ge \cdots \ge b_{\pi(n)}$.

This is easily proved. If $j < k$ and $b_{\pi(j)} > b_{\pi(k)}$ then by switching the values of $\pi(j)$ and $\pi(k)$, the change in Σ is

$$a_jb_{\pi(k)} + a_kb_{\pi(j)} - (a_jb_{\pi(j)} + a_kb_{\pi(k)}) = (a_k - a_j)(b_{\pi(j)} - b_{\pi(k)}) \ge 0.$$

It follows that Σ is maximum when $b_{\pi(1)}, b_{\pi(2)}, \ldots, b_{\pi(n)}$ are in non-decreasing order. In the same way, Σ is mimimum when $b_{\pi(1)} \ge b_{\pi(2)} \ge \cdots \ge b_{\pi(n)}$.

For the special case where $0 < a_1 \le a_2 \le \cdots \le a_n$ and $b_k = 1/a_k$ for $k = 1, 2, \ldots, n$, we have

$$n = \frac{a_1}{a_1} + \frac{a_2}{a_2} + \cdots + \frac{a_n}{a_n} \le \frac{a_1}{a_{\pi(1)}} + \frac{a_2}{a_{\pi(2)}} + \cdots + \frac{a_n}{a_{\pi(n)}}$$

for any permutation π. As an application of this result, let us give another proof of the AM-GM inequality. Suppose $0 < x_1 \le x_2 \le \cdots \le x_n$ and let

$$G = \sqrt[n]{x_1 x_2 \cdots x_n}.$$

Set

$$a_k = \frac{x_1 x_2 \cdots x_k}{G^k}, \quad k = 1, 2, \ldots, n.$$

Then $a_n = 1$ and

$$n \le \frac{a_1}{a_n} + \frac{a_2}{a_1} + \cdots + \frac{a_n}{a_{n-1}} = \frac{x_1}{G} + \frac{x_2}{G} + \cdots + \frac{x_n}{G}.$$

Thus we have proved the AM-GM inequality

$$\sqrt[n]{x_1 x_2 \cdots x_n} \le \frac{x_1 + x_2 + \cdots + x_n}{n}$$

one more time.

The next problem was on the 1981 USAMO. The elegant proof given here had appeared earlier in the literature, and is due to Professor Andrew Gleason.

Example 2.46 (1981 USAMO) *If x is a positive real number and n is a positive integer, prove that*

$$\sum_{k=1}^{n} \frac{\lfloor kx \rfloor}{k} \le \lfloor nx \rfloor,$$

where $\lfloor t \rfloor$ denotes the greatest integer less than or equal to t.

Solution. First note that for any integer q, we have $\lfloor n(q + x) \rfloor = nq + \lfloor nx \rfloor$ and

$$\sum_{k=1}^{n} \frac{\lfloor k(q + x) \rfloor}{k} = nq + \sum_{k=1}^{n} \frac{\lfloor kx \rfloor}{k},$$

so it suffices to prove the given equality for $0 \le x < 1$. Secondly, the left-hand side of the inequality has "jumps" at rational points $x = a/b$ where $1 \le a < b \le n$ and is constant in between, so it suffices to prove that

$$\sum_{k=1}^{n} \frac{\lfloor ka/b \rfloor}{k} \le \lfloor na/b \rfloor, \qquad 1 \le a < b \le n, \ (a, b) = 1.$$

By the Division Algorithm, for $k = 1, 2, \ldots, n$ there are integers q_k, r_k such that

$$ka = q_k b + r_k, \qquad 0 \le r_k < b, \ k = 1, 2, \ldots, n.$$

Substituting $\lfloor ka/b \rfloor = q_k = ka - r_k$ $(k = 1, 2, \ldots, n)$, the inequality to be proved becomes

$$\sum_{k=1}^{n} \frac{r_k}{k} \ge r_n.$$

Since $r_k \ge 0$ for every k and $r_n < b$, it suffices to prove that

$$\sum_{k=1}^{b-1} \frac{r_k}{k} \ge b - 1.$$

We claim that $(r_1, r_2, \ldots, r_{b-1})$ is a permutation of $(1, 2, \ldots, b-1)$. To verify this claim, first note that it is impossible that $r_k = 0$ for some $1 \le k < b$ since this gives $a/b = q_k/k$, whereas a/b is irreducible. Also $r_j = r_k$ for some $1 \le j < k < b$ is impossible since this gives $a/b = (q_k - q_j)/(k - j)$, again contradicting the fact that a/b is irreducible. It follows that $(r_1, r_2, \ldots, r_{b-1})$ is a permutation of $(1, 2, \ldots, b-1)$ as claimed, and thus

$$\frac{r_1}{1} + \frac{r_2}{2} + \cdots + \frac{r_{b-1}}{b-1} \ge \frac{1}{1} + \frac{2}{2} + \cdots + \frac{b-1}{b-1} = b - 1$$

by the rearrangement inequality. □

Theorem 2.19 (Chebyshev's inequality) *Suppose* $x_1 \le x_2 \le \cdots \le x_n$ *and* $y_1 \le y_2 \le \cdots \le y_n$, *or* $x_1 \ge x_2 \ge \cdots \ge x_n$ *and* $y_1 \ge y_2 \ge \cdots \ge y_n$ *(in short, the x's and y's are similarly ordered). If* $\lambda_1, \lambda_2, \ldots, \lambda_n > 0$ *and* $\sum_{i=1}^{n} \lambda_i = 1$ *then*

$$\left(\sum_{i=1}^{n} \lambda_i x_i\right)\left(\sum_{i=1}^{n} \lambda_i y_i\right) \le \sum_{i=1}^{n} \lambda_i x_i y_i.$$

Since the x's and y's are similarly ordered, we have $(x_i - x_j)(y_i - y_j) \ge 0$ for all i, j. Thus

$$\sum_{i,j=1}^{n} \lambda_i \lambda_j (x_i - x_j)(y_i - y_j) \ge 0.$$

Multiplied out, this gives Chebyshev's inequality. Note that as a special case, we have

$$\left(\frac{1}{n}\sum_{i=1}^{n} x_i\right)\left(\frac{1}{n}\sum_{i=1}^{n} y_i\right) \le \frac{1}{n}\sum_{i=1}^{n} x_i y_i,$$

or

$$\sum_{i-1}^{n} x_i y_i \ge \frac{1}{n}\left(\sum_{i=1}^{n} x_i\right)\left(\sum_{i=1}^{n} y_i\right).$$

Example 2.47 (Proposed for the 1990 IMO) *Let a, b, c, d be non-negative real numbers such that $ab + bc + cd + da = 1$. Show that*

$$\frac{a^3}{b+c+d} + \frac{b^3}{a+c+d} + \frac{c^3}{a+b+d} + \frac{d^3}{a+b+c} \ge \frac{1}{3}.$$

Solution. Let A, B, C, D denote $b+c+d$, $a+c+d$, $a+b+d$, $a+b+c$, respectively. From $(a-b)^2 + (b-c)^2 + (c-d)^2 + (d-a)^2 \ge 0$ we have

$$a^2 + b^2 + c^2 + d^2 \ge ab + bc + cd + da = 1.$$

By symmetry, we may assume that $a \ge b \ge c \ge d$. Then a^3, b^3, c^3, d^3 and $\frac{1}{A}, \frac{1}{B}, \frac{1}{C}, \frac{1}{D}$ are similarly ordered, so by Chebyshev's inequality,

$$\frac{a^3}{A} + \frac{b^3}{B} + \frac{c^3}{C} + \frac{d^3}{D} \ge \frac{1}{4}(a^3 + b^3 + c^3 + d^3)\left(\frac{1}{A} + \frac{1}{B} + \frac{1}{C} + \frac{1}{D}\right).$$

Also by Chebyshev's inequality,

$$a^3 + b^3 + c^3 + d^3 \ge \frac{1}{4}(a^2 + b^2 + c^2 + d^2)(a + b + c + d).$$

Since $a^2 + b^2 + c^2 + d^2 \ge 1$ and $3(a + b + c + d) = A + B + C + D$, we have

$$\frac{a^3}{A} + \frac{b^3}{B} + \frac{c^3}{C} + \frac{d^3}{D} \ge \frac{1}{48}(A + B + C + D)\left(\frac{1}{A} + \frac{1}{B} + \frac{1}{C} + \frac{1}{D}\right).$$

Finally, by (2.7)

$$\frac{a^3}{A} + \frac{b^3}{B} + \frac{c^3}{C} + \frac{d^3}{D} \ge \frac{1}{3}.$$

Equality holds if and only if $a = b = c = d$. □

Proving inequalities is not simply a matter of knowing a collection of theorems. Intuition must guide the choice of facts used in the proof. The last six examples are illustrations of olympiad-level inequality problems. The classical inequalities used in solving them are very basic, but some insight is needed to choose the right approach.

Example 2.48 (Proposed for the 1992 IMO) *Prove that if* $x, y, z > 1$ *and* $\frac{1}{x} + \frac{1}{y} + \frac{1}{z} = 2$ *then*

$$\sqrt{x+y+z} \geq \sqrt{x-1} + \sqrt{y-1} + \sqrt{z-1}.$$

Solution. Set $x = 1 + u^2$, $y = 1 + v^2$, $z = 1 + w^2$ where $u, v, w > 0$. Then it is enough to prove

$$(3 + u^2 + v^2 + w^2) \geq (u + v + w)^2,$$

where

$$\frac{1}{1+u^2} + \frac{1}{1+v^2} + \frac{1}{1+w^2} = 2.$$

Simplifying, we find that the problem reduces to proving

$$uv + uw + vw \leq \frac{3}{2} \quad \text{where} \quad u^2v^2 + u^2w^2 + v^2w^2 + 2u^2v^2w^2 = 1.$$

Set $p = uv + uw$, $q = uv - uw$, and $r = vw$. The inequality to be proved is

$$p + r \leq \frac{3}{2} \quad \text{where} \quad \frac{p^2+q^2}{2} + r^2 + \frac{(p^2-q^2)r}{2} = 1.$$

Clearly $r < 1$, so we have

$$(1 + r)p^2 = 2(1 - r^2) - (1 - r)q^2 \leq 2(1 - r^2),$$

and thus $p \leq \sqrt{2(1-r)}$. Hence

$$p + r \leq \sqrt{2(1-r)} + r = \frac{3}{2} - \left(\sqrt{1-r} - \frac{\sqrt{2}}{2}\right)^2 \leq \frac{3}{2}.$$

Looking back at the steps of the argument, it is clear that equality holds if and only if $p = 1$, $q = 0$, and $r = 1/2$. This translates to $uv = uw = vw = 1/2$, so $u^2 = v^2 = w^2 = 1/2$ and thus $x = y = x = 3/2$. □

Example 2.49 (1984 IMO) *Prove that*

$$0 \le xy + yz + zx - 2xyz \le 7/27$$

where x, y, and z are nonnegative real numbers for which $x + y + z = 1$.

Solution. By symmetry, we may assume that $x \le y \le z$. Thus $x \le 1/3$, and using $x + y + z = 1$, we obtain the lower bound

$$xy + yz + zx - 2xyz = x(1 - x) + yz(1 - 2x) \ge 0.$$

To prove the upper bound, note that the AM-GM inequality yields $yz \le (1 - x)^2/4$, so the desired inequality follows if we can show

$$x(1 - x) + \left(\frac{1 - x}{2}\right)^2 (1 - 2x) \le \frac{7}{27}, \qquad 0 \le x \le 1/3.$$

Simplifying this inequality a bit, we find that it can be written

$$x^2(1 - 2x) \le \frac{1}{27}, \qquad 0 \le x \le 1/3.$$

With another application of AM-GM, the solution is complete:

$$x^2(1 - 2x) \le \left(\frac{x + x + (1 - 2x)}{3}\right)^3 = \frac{1}{27}.$$

Equality holds if and only if $x = y = z = 1/3$. □

Example 2.50 (1987 Yugoslav Math Olympiad) *Prove that for all nonnegative numbers a, b,*

$$\frac{1}{2}(a + b)^2 + \frac{1}{4}(a + b) \ge a\sqrt{b} + b\sqrt{a}.$$

Solution. Since the left side of the inequality depends only on the sum $a + b$, it seems natural to begin by finding the greatest value of the right side for a fixed value of $a + b$. A reasonable guess is that the maximum is attained when $a = b = (a + b)/2$. Thus we want to prove that

$$a\sqrt{b} + b\sqrt{a} \le (a + b)\sqrt{(a + b)/2}$$

for all nonnegative a, b. By the AM-GM inequality, we have $\sqrt{ab} \le (a + b)/2$. Also, using the same inequality or simply noting $(\sqrt{a} - \sqrt{b})^2 + (\sqrt{a} - \sqrt{b})^2 = 2(a + b)$, we have $\sqrt{a} + \sqrt{b} \le \sqrt{2(a + b)}$.

Hence

$$a\sqrt{b} + b\sqrt{a} = \sqrt{ab}(\sqrt{a} + \sqrt{b}) \leq \frac{a+b}{2}\sqrt{2(a+b)}$$

as required, with equality when $a = b$. Let $c = a + b$. Now the solution follows if we show that

$$\frac{c^2 + c/2}{2} \geq \sqrt{c^3/2}$$

for nonnegative c. Again, AM-GM does the trick, and further shows that equality holds if and only if $c = 0$ or $c = 1/2$. Thus we have

$$\frac{1}{2}(a+b)^2 + \frac{1}{4}(a+b) \geq a\sqrt{b} + b\sqrt{a},$$

with equality if and only if $a = b = 0$ or $a = b = 1/4$. □

Example 2.51 (Proposed for the 1987 IMO) *Prove that if x, y, z are real numbers such that $x^2 + y^2 + z^2 = 2$ then*

$$x + y + z \leq 2 + xyz.$$

Solution. From $(x-y)^2 + z^2 \geq 0$ and $x^2 + y^2 + z^2 = 2$ we have $xy \leq 1$. In like manner, $xz \leq 1$ and $yz \leq 1$. Hence

$$2(1 - xy)(1 - xz)(1 - yz) + (xyz)^2 \geq 0,$$

which gives

$$2(xy + xz + yz) - 2(x + y + z)xyz + (xyz)^2 \leq 2.$$

Using this result together with $x^2 + y^2 + z^2 = 2$, we obtain

$$0 \leq (x + y + z - xyz)^2 =$$

$$(x^2 + y^2 + z^2) + 2(xy + xz + yz) - 2(x + y + z)xyz + (xyz)^2 \leq 4,$$

and thus

$$-2 \leq x + y + z - xyz \leq 2.$$

Equality holds in the original inequality ($x + y + z \leq 2 + xyz$) for $(x, y, z) = (1,1,0), (1,0,1), (0,1,1)$. □

Example 2.52 (Proposed for the 1993 IMO) *Prove that*

$$\frac{a}{b+2c+3d} + \frac{b}{c+2d+3a} + \frac{c}{d+2a+3b} + \frac{d}{a+2b+3c} \geq \frac{2}{3}$$

for all positive numbers a, b, c, d.

Solution. By Cauchy's inequality, if $x_1, x_2, \ldots, x_n$ and $y_1, y_2, \ldots, y_n$ are positive numbers, then

$$\left(\sum_{i=1}^{n} \frac{x_i}{y_i}\right)\left(\sum_{i=1}^{n} x_i y_i\right) \geq \left(\sum_{i=1}^{n} x_i\right)^2.$$

Apply this in the case of $x_1 = a$, $y_1 = b + 3c + 2d$, and so on, and note that $a(b + 2c + d) + b(c + 2d + 3a) + c(d + 2a + 3b) + d(a + 2b + 3c) = 4(ab + ac + ad + bc + bd + cd) \leq \frac{3}{2}(a + b + c + d)^2$, where the last inequality follows from $8(ab + ac + ad + bc + bd) = 3(a + b + c + d)^2 - [(a-b)^2 + (a-c)^2 + (a-d)^2 + (b-c)^2 + (b-d)^2]$. Thus we obtain

$$\frac{a}{b + 2c + 3d} + \frac{b}{c + 2d + 3a} + \frac{c}{d + 2a + 3b} + \frac{d}{a + 2b + 3c} \geq \frac{2}{3}.$$

Equality holds if and only if $a = b = c = d$. □

Example 2.53 (1995 IMO) *Let a, b, c be positive real numbers such that $abc = 1$. Prove that*

$$\frac{1}{a^3(b + c)} + \frac{1}{b^3(c + a)} + \frac{1}{c(a + b)} \geq \frac{3}{2}.$$

Solution. Let $x = 1/a$, $y = 1/b$, and $z = 1/c$. Then

$$\frac{1}{a^3(b + c)} + \frac{1}{b^3(c + a)} + \frac{1}{c(a + b)} = \frac{x^2}{y + z} + \frac{y^2}{x + z} + \frac{z^2}{x + y}$$

and $xyz = 1$. Let

$$(u_1, u_2, u_3) = \left(\frac{x}{\sqrt{y + z}}, \frac{y}{\sqrt{x + z}}, \frac{z}{\sqrt{x + y}}\right)$$

and

$$(v_1, v_2, v_3) = (\sqrt{y + z}, \sqrt{x + z}, \sqrt{x + y}).$$

By means of Cauchy's inequality and AM-GM, we obtain

$$\begin{aligned}\frac{x^2}{y + z} + \frac{y^2}{x + z} + \frac{z^2}{x + y} &= u_1^2 + u_2^2 + u_3^2 \\ &\geq \frac{(u_1 v_1 + u_2 v_2 + u_3 v_3)^2}{v_1^2 + v_2^2 + v_3^2}\end{aligned}$$

$$= \frac{x+y+z}{2}$$
$$\geq \frac{3}{2}\sqrt[3]{xyz}$$
$$= \frac{3}{2}.$$

Equality holds if and only if $a = b = c = 1$. □

Exercises for Section 2.4

1. Show that $a^6 - a^5 + a^4 + a^2 - a + 1 > 0$ for every real number a.
2. Show that if a, b, c, are positive real numbers, then
$$\frac{a^2}{b^2} + \frac{b^2}{c^2} + \frac{c^2}{a^2} \geq \frac{b}{a} + \frac{c}{b} + \frac{a}{c}.$$
3. Show that if $a, b > 0$ and $a + b = 1$, then
$$\left(a + \frac{1}{a}\right)^2 + \left(b + \frac{1}{b}\right)^2 \geq \frac{25}{2}.$$
4. Let a, b, c be positive real numbers. Prove that
$$\frac{1}{a} + \frac{1}{b} + \frac{1}{c} \leq \frac{a^8 + b^8 + c^8}{a^3b^3c^3}.$$
5. Let a, b, c be real numbers between 0 and 1. Prove that not all of the numbers $a(1-b), b(1-c), c(1-a)$ can be greater than $\frac{1}{4}$.
6. Let $a_1, a_2, \ldots, a_n$ be positive real numbers and set $S = a_1 + a_2 + \cdots + a_n$. Prove that
$$\frac{S}{S-a_1} + \frac{S}{S-a_2} + \cdots + \frac{S}{S-a_n} \geq \frac{n^2}{n-1}.$$
7. Show that if a, b, c are positive real numbers, then
$$\frac{a}{b+c} + \frac{b}{a+c} + \frac{c}{a+b} \geq \frac{3}{2}.$$
8. The area of a triangle with sides of length a, b, c is
$$A = \sqrt{s(s-a)(s-b)(s-c)},$$
where $s = (a+b+c)/2$ is the **semiperimeter**. Using this formula and the AM-GM inequality, show that the area of a triangle with semiperimeter s is at most $\sqrt{3}s^2/9$.

9. Using the AM-GM inequality, find the largest possible volume of a right circular cylinder that is inscribed in a sphere of radius R. *Hint:* Maximize $V = \pi r^2 h$ subject to $r^2 + (h/2)^2 = R^2$.

10. Let a, b, c be any three real numbers. Show that the simultaneous equations

$$a^2 + b^2 + c^2 + 3(x^2 + y^2 + z^2) = 6,$$

$$ax + by + cz = 4,$$

have no real solutions.

11. Prove that if a, b, c are arbitrary real numbers, then

$$a^4(1 + b^4) + b^4(1 + c^4) + c^4(1 + a^4) \geq 6a^2b^2c^2.$$

When is there equality?

12. Prove the AM-GM inequality by carrying out the following program. (i) Given that the theorem is true for $n = 2$, in other words, $(a_1 + a_2)/2 \geq \sqrt{a_1a_2}$ for arbitrary nonnegative numbers a_1 and a_2, use mathematical induction to prove that the theorem is true whenever n is a power of two. (ii) Prove that if the theorem is true for n it is also true for all $m < n$. *Hint:* Given $a_1, a_2, \ldots, a_m$, apply AM-GM to the n numbers $a_1, \ldots, a_m, \overline{a}, \ldots, \overline{a}$, where the last $n - m$ terms are

$$\overline{a} = \frac{a_1 + a_2 + \cdots + a_m}{m},$$

and note that the arithmetic mean of these n numbers is also $\overline{a}$. Thus

$$a_1a_2 \cdots a_m\overline{a}^{n-m} \leq \overline{a}^n,$$

and we may cancel $\overline{a}^n$ to get

$$a_1a_2 \cdots a_m \leq \left(\frac{a_1 + a_2 + \cdots + a_m}{m}\right)^m.$$

This trick of going from the truth of statement n to that of $m < n$ is sometimes called "backward induction." This ingenious strategy is due to Cauchy. The same method can be used to prove Jensen's inequality.

13. Prove Cauchy's inequality. *Hint:* First note that the inequality is clearly true if all b_i are 0. Now assume that

$$b_1^2 + \cdots + b_n^2 > 0$$

and consider the quadratic polynomial

$$P(x) = (a_1 - b_1 x)^2 + (a_2 - b_2 x)^2 + \cdots + (a_n - b_n x)^2.$$

Note that $P(x) \geq 0$ for all real x. In particular, $P(x) \geq 0$ for

$$x = \frac{a_1 b_1 + \cdots + a_n b_n}{b_1^2 + \cdots + b_n^2},$$

and this gives the desired conclusion.

14. Given n positive numbers $a_1, a_2, \ldots, a_n$, let σ_k denote the corresponding kth elementary symmetric function and define **symmetric means** $\Sigma_1, \Sigma_2, \ldots, \Sigma_n$ by

$$\Sigma_k = \left\{ \binom{n}{k}^{-1} \sigma_k \right\}^{1/k}.$$

The **symmetric mean inequality** states that

$$\Sigma_1 \geq \Sigma_2 \geq \cdots \geq \Sigma_n.$$

For the case of three variables, the symmetric means are

$$\Sigma_1 = \frac{a + b + c}{3}, \quad \Sigma_2 = \sqrt{\frac{ab + bc + ca}{3}}, \quad \Sigma_3 = \sqrt[3]{abc}.$$

Prove the symmetric mean inequality for the case $n = 3$.

15. Prove that if $a_1, a_2, \ldots, a_n$ are positive real numbers whose product is 1, then $(1 + a_1)(1 + a_2) \cdots (1 + a_n) \geq 2^n$.

Olympiad Problems for Chapter 2

1. If a and b are two roots of $x^4 + x^3 - 1 = 0$, prove that ab is a root of $x^6 + x^4 + x^3 - x^2 - 1 = 0$. [1977 USAMO]

2. The product of two of the four roots of the quartic equation

$$x^4 - 18x^3 + kx^2 + 200x - 1984 = 0$$

is -32. Determine the value of k. [1984 USAMO]

3. Let x_1, x_2, x_3 be the roots of $x^3 - 6x^2 + ax + a = 0$. Determine all real numbers a such that

$$(x_1 - 1)^3 + (x_2 - 2)^3 + (x_3 - 3)^3 = 0.$$

Also, for each such a, determine the corresponding values of x_1, x_2, x_3. [1983 Austrian Olympiad]

4. If $P(x)$ denotes a polynomial of degree n such that $P(k) = k/(k+1)$ for $k = 0, 1, 2, \ldots, n$, determine $P(n+1)$. [1975 USAMO]

5. Find all real roots of the equation $\sqrt{x^2 - p} + 2\sqrt{x^2 - 1} = x$, where p is a real parameter. [1963 IMO]

6. Determine all real numbers x that satisfy the inequality

$$\sqrt{3 - x} - \sqrt{x + 1} > \frac{1}{2}.$$

[1962 IMO]

7. Prove that the roots of

$$x^5 + ax^4 + bx^3 + cx^2 + dx + e = 0$$

cannot *all* be real if $2a^2 < 5b$. [1983 USAMO]

8. Let a and b be real numbers for which the equation

$$x^4 + ax^3 + bx^2 + ax + 1 = 0$$

has at least one real solution. For all such pairs (a, b), find the minimum value of $a^2 + b^2$. [1973 IMO]

9. If a, b, c, d, e are positive numbers bounded by p and q, specifically, if $0 < p \leq a, b, c, d, e \leq q$, prove that

$$(a + b + c + d + e)\left(\frac{1}{a} + \frac{1}{b} + \frac{1}{c} + \frac{1}{d} + \frac{1}{e}\right)$$

does not exceed $25 + 6(\sqrt{p/q} - \sqrt{q/p})^2$ and determine when there is equality. [1977 USAMO]

10. Prove that the inequality

$$\sum_{n=1}^{r}\sum_{m=1}^{r} \frac{a_m a_n}{m + n} \geq 0$$

holds for any real numbers $a_1, a_2, \ldots, a_r$. Find conditions for equality. [1991-92 Polish Olympiad]

CHAPTER 3 Combinatorics

3.1 What is Combinatorics?

In 1666 Gottfried Wilhelm Leibnitz, then only twenty years old, wrote *Ars Combinatoria*, in which he used the word "combinatorial" to describe an area of mathematics for which he could forsee many wonderful uses, from fundamentals of logic to such practical matters as the construction of locks. Leibnitz was young and inexperienced at the time, so much of the mathematics in *Ars Combinatoria* wasn't new. Nevertheless, the scope of his approach is impressive. He saw "applications to the whole sphere of sciences" and listed military science, grammar, law, and medicine as areas where combinatorics might be applied. In recent years, the vision of Leibnitz has become a reality, and combinatorics is now a foundational subject for computer science and is used in many creative ways in science and technology.

Combinatorics is concerned with configurations involving finite sets. Its basic questions involve existence, construction, enumeration (counting), and extremal properties. Among the common objects of concern in combinatorics are permutations, 0–1 sequences, partitions, matrices, graphs, and set systems (hypergraphs). Problems in combinatorics are often both attractive and difficult. Techniques used to solve problems in combinatorics include bijections, recurrence relations, generating functions, the inclusion-exclusion principle, the pigeonhole principle, and mathematical induction.

3.2 Basics of Counting

We shall denote the number of elements of a finite set S by $|S|$. Our basic problem is easy to state. Given a set S of interest, what is $|S|$? There are two basic rules that are useful in dealing with this problem. Using these two rules, we can break down a complicated counting problem into simple subproblems.

Rule 3.1 (Sum) *If* $A \cap B = \emptyset$ *then* $|A \cup B| = |A| + |B|$.

The second rule is expressed in terms of the Cartesian product $A \times B = \{(a, b) | \, a \in A, b \in B\}$.

Rule 3.2 (Product) $|A \times B| = |A| \cdot |B|$.

The Sum Rule generalizes to n sets: if $A_1, A_2, \ldots, A_n$ are pairwise disjoint sets, then

$$|A_1 \cup A_2 \cup \cdots \cup A_n| = |A_1| + |A_2| + \cdots + |A_n|.$$

The Product Rule generalizes to

$$|A_1 \times A_2 \times \cdots \times A_n| = |A_1| \cdot |A_2| \cdots |A_n|.$$

More generally still, if a collection of n-tuples $(a_1, a_2, \ldots, a_n)$ is formed by a sequence of choices such that for any possible choice of $(a_1, a_2, \ldots, a_{k-1})$, the number of available choices for a_k is N_k, independent of the choices made thus far, then the total number of n-tuples that can be formed is the product $N_1 N_2 \cdots N_n$.

Our standard example of an n-element set is $\{1, 2, \ldots, n\}$, which we shall denote by $[n]$. By the Product Rule, the number of permutations of $[n]$ is $n \cdot (n-1) \cdots 2 \cdot 1 = n!$. By the same argument, there are $n(n-1) \cdots (n-k+1)$ k-tuples of elements from $[n]$. To count unordered selections (subsets), we divide by $k!$ and get the familiar result that the number of k-element subsets of $[n]$ is $n(n-1) \cdots (n-k+1)/k! = \binom{n}{k}$. The Product Rule is often used in this way. On purpose, we first count a larger class of configurations, and then compensate for the overcount by dividing by the appropriate factor.

Example 3.1 *How many ways are there to distribute the elements of* $[n]$ *among* k *unlabeled boxes if* a_1 *of the boxes contain just one element,* a_2 *contain two elements, and so on, and no box is empty?*

Solution. Such an arrangement is called a **partition** of $[n]$ into k **blocks**; the n-tuple $(a_1, a_2, \ldots, a_n)$ specifies the **type** of the partition. Note that the nonnegative integers $a_1, a_2, \ldots, a_n$ must satisfy $a_1 + 2a_2 + 3a_3 + \cdots + na_n = n$ and $a_1 + a_2 + a_3 + \cdots + a_n = k$. In order to count the number of partitions of $[n]$ of type $(a_1, a_2, \ldots, a_n)$, imagine first a different situation in which the boxes are labeled, and that a box containing j elements has j separate compartments where individual elements can be placed. In this case, there are n distinguished positions for the elements of $[n]$ and thus $n!$ possible arrangements. Now we must compensate for the fact that the boxes are not labeled and do not have such compartments. For each j any permutation of the a_j labeled boxes containing j elements gives the same partition hence we need to divide the previous count by $a_1!a_2! \cdots a_n!$. Also, we need to divide by $(1!)^{a_1}(2!)^{a_2} \cdots (n!)^{a_n}$ because permutations of elements within a box give the same partition. Hence we arrive at the desired formula for the number of partitions of type $(a_1, a_2, \ldots, a_n)$, namely

$$P(a_1, a_2, \ldots, a_n; n) = \frac{n!}{a_1!a_2! \cdots a_n!(1!)^{a_1}(2!)^{a_2} \cdots (n!)^{a_n}}. \quad \square$$

The most natural and attractive way to solve a new counting problem is to show it can be reduced to one we have already solved.

Principle 3.1 (Bijection) *Suppose there exists a function* $f : A \to B$ *such that (i)* $f(a_1) = f(a_2)$ *implies* $a_1 = a_2$ *and (ii) for each* $b \in B$ *there is some* $a \in A$ *such that* $f(a) = b$. *Then* $|A| = |B|$.

A function that that satisfies (i) and (ii) is called a **bijection** (or a **one-to-one correspondence**), and an argument that shows that $|A| = |B|$ by exhibiting such a function is called a **bijective** or **combinatorial proof**. In practice, the proof that a given mapping f is a bijection is usually carried out by exhibiting a function $g = f^{-1}$

satisfying $g(f(a)) = a$ for all $a \in A$ and $f(g(b)) = b$ for all $b \in B$. Several examples follow.

Example 3.2 *Show that the number of subsets of* $[2n]$ *having the same number of even elements as odd elements is* $\binom{2n}{n}$.

Solution. Given $X \subseteq [2n]$ with k even elements and k odd elements, replace the k odd elements of X by the $n - k$ odd elements not in X to obtain $f(X) \subseteq [2n]$ with k even elements and $n - k$ odd elements. Thus f maps the given collection of subsets of $[2n]$ to the collection of all n-element subsets of $[2n]$. It is easy to see that f is a bijection, with $f^{-1}(Y)$ given by exchanging the odd elements of Y for the odd elements of $[2n]$ not in Y. □

Example 3.3 *Find the number of k-tuples* $(S_1, S_2, \ldots, S_k)$ *satisfying*

$$S_1 \subseteq S_2 \subseteq \cdots \subseteq S_k \subseteq [n].$$

Solution. The desired number is $(k + 1)^n$. To see this, we construct a bijection that associates with each set sequence $(S_1, S_2, \ldots, S_k)$ satisfying

$$S_1 \subseteq S_2 \subseteq \cdots \subseteq S_k \subseteq [n]$$

a corresponding "codeword" $(a_1, a_2, \ldots, a_n)$ in which each a_j assumes one of $k + 1$ possible values. To construct the codeword that goes with a given nested sequence of sets, we proceed as follows. For each j between 1 and n, if $j \in S_k$ let a_j be the index of the first set in the chain $S_1, S_2, \ldots, S_k$ that contains j. If $j \notin S_k$, set $a_j = k + 1$. To see that this mapping is a bijection, note that the chain $S_1, S_2, \ldots, S_k$ is uniquely constructed from its codeword by setting

$$S_i = \{j \mid a_j \leq i\}, \qquad i = 1, 2, \ldots, k.$$

Thus there are just as many chains as there are codewords, namely $(k + 1)^n$. □

A **0–1 string** of length n is an n-tuple $(a_1, a_2, \ldots, a_n)$ in which each element is either 0 or 1.

Example 3.4 *Show that the number of 0–1 strings of length n with exactly m zeros that are followed immediately by ones is* $\binom{n+1}{2m+1}$.

Solution. Given a 0–1 string $(a_1, a_2, \ldots, a_n)$, enlarge it to a string of length $n + 2$ by adding $a_0 = 1$ and $a_{n+1} = 0$. Then associate with the enlarged string a corresponding one $(b_1, b_2, \ldots, b_{n+1})$ by setting $b_i = 0$ if $a_i = a_{i-1}$ and $b_i = 1$ if $a_i \neq a_{i-1}$. It is easy to check that the mapping just defined is a bijection from the set of 0–1 strings $(a_1, a_2, \ldots, a_n)$ that have exactly m zeros that are followed immediately by ones to the set of 0–1 strings $(b_1, b_2, \ldots, b_{n+1})$ that have $2m + 1$ ones. There are $\binom{n+1}{2m+1}$ strings in the latter set, so the proof is complete. □

Example 3.5 *How many k-element subset of [n] contain no two consecutive integers?*

Solution. For $\{a_1, a_2, \ldots, a_k\} \subset [n]$ where $a_1 < a_2 < \cdots < a_k$ and no two of the a_i's are consecutive, let $\{b_1, b_2, \ldots b_k\}$ be the subset of $[n-k+1]$ given by setting $b_i = a_i - i + 1$, $i = 1, 2, \ldots, k$. To construct $\{a_1, a_2, \ldots, a_k\}$ from $\{b_1, b_2, \ldots, b_k\}$ set $a_i = b_i + i - 1$, $i = 1, 2, \ldots, k$. Thus we have a bijection from the k-element subsets of $[n]$ containing no two consecutive integers to the k-element subsets of $[n-k+1]$ It follows that the number of k-element subsets of $[n]$ containing no two consecutive integers is $\binom{n-k+1}{k}$. □

Example 3.6 (Ballot Problem) *How many A-B sequences consisting of n A's and n B's have the property that, when read from left to right, the number of A's never lags behind the number of B's? (Think of an election in which the votes are counted one at a time; at the end there is a tie, but candidate A never trails candidate B as the votes are counted.)*

Solution. We shall refer to the fact that A never trails B as the votes are counted as the **ballot condition.** Without the ballot condition, there are $\binom{2n}{n}$ possible sequences. Of these sequences, certain ones are "bad" (A trails B at some point). In a bad sequence, there is a first time where the number of B's exceeds the number of A's. At this point, there is one more B than A, and reversing all symbols up to that point gives a sequence with $n+1$ A's and $n-1$ B's. This "reflection" (exchanging A's and B's) provides a bijection from the set of bad sequences onto the set of all sequences with $n+1$ A's and $n-1$ B's. This shows the number of bad sequences is $\binom{2n}{n+1}$, so there are

$$\binom{2n}{n} - \binom{2n}{n+1} = \left[1 - \frac{n}{n+1}\right]\binom{2n}{n} = \frac{1}{n+1}\binom{2n}{n}$$

acceptable ones. □

The number

$$C_n = \frac{1}{n+1}\binom{2n}{n}$$

is called the nth **Catalan number**. The first six Catalan numbers are $C_0 = 1$, $C_1 = 1$, $C_2 = 2$, $C_3 = 5$, $C_4 = 14$, $C_5 = 42$.

Note. The above argument is due to Antoine Désiré André; its essential idea is called the **reflection principle**. The following slight generalization is solved by the same technique. Suppose that A gets $k \geq m/2$ out of m votes and never trails as the votes are counted. Then A leads by $2k-m$ votes at the end, and out of a total of $\binom{m}{k}$ voting sequences, $\binom{m}{k+1}$ violate the ballot condition. Thus the number of acceptable sequences is

$$\binom{m}{k} - \binom{m}{k+1} = \frac{2k-m+1}{k+1}\binom{m}{k}.$$

This result can be interpreted in terms of **random walks.** The number of random walks on $\mathbb{Z}$, starting at the origin and ending at $x = a$ after $n = a + 2r$ steps, never having visited a point to the left of the origin, is given by

$$\frac{2(a+r)-(a+2r)+1}{a+r+1}\binom{a+2r}{a+r} = \frac{a+1}{a+r+1}\binom{n}{r}.$$

As we shall see, Catalan numbers show up in many important combinatorial problems. Two examples follow.

Example 3.7 *Find the number of nondecreasing functions $f : [n] \to [n]$ that satisfy $f(k) \leq k$ for $k = 1, 2, \ldots, n$.*

Solution. Let us call such a function *admissible*. It is convenient to represent the function by an n-tuple $(f(1), f(2), \ldots, f(n))$. For $n = 1, 2, 3$, and 4, the admissible functions are thus represented as follows:

$$\begin{array}{lllll} n = 1: & (1) & & & \\ n = 2: & (1,1) & (1,2) & & \\ n = 3: & (1,1,1) & (1,1,2) & (1,1,3) & (1,2,2) \\ & (1,2,3) & & & \\ n = 4: & (1,1,1,1) & (1,1,1,2) & (1,1,1,3) & (1,1,1,4) \\ & (1,1,2,2) & (1,1,2,3) & (1,1,2,4) & (1,1,3,3) \\ & (1,1,3,4) & (1,2,2,2) & (1,2,2,3) & (1,2,2,4) \\ & (1,2,3,3) & (1,2,3,4). & & \end{array}$$

There are 1, 2, 5, and 14 admissible functions for $n = 1, 2, 3$ and 4 respectively, so we suspect the solution of our problem will involve Catalan numbers. To prove that there are

$$C_n = \frac{1}{n+1}\binom{2n}{n}$$

admissible functions from $[n]$ into $[n]$, we look for a bijection from the set of all sequences satisfying the ballot condition (see Example 3.6) onto the set of admissible functions. We shall hereafter refer to sequences satisfying the ballot condition as **ballot sequences**. Given a ballot sequence, for $k = 1, 2, \ldots, n$ let $p(k)$ signify the position of the kth A in the sequence and let $m(k)$ be the margin by which A leads B just before the kth A appears. Set $f(k) = k - m(k)$. Note that $f(k) - 1$ is the number of B's in positions 1 through $p(k)$, so f is nondecreasing. The fact that A never lags behind B implies $m(k) \geq 0$, so $f(k) \leq k$. Thus f is admissible. Note that given any admissible function f, we can construct the associated ballot sequence by making the first term an A, placing $f(k) - f(k-1)$ B's between the $(k-1)$st A and the kth A for $k = 2, 3, \ldots, n$, and finishing the sequence with a suitable string of B's. It follows that the mapping from ballot sequences to admissible functions is a bijection. □

Example 3.8 *Show that the number of permutations π of $[n]$ with no triple $i < j < k$ such that $\pi(i) > \pi(k) > \pi(j)$ is C_n, the nth Catalan number.*

Solution. The following argument uses a concept from computer science. A **stack** is a device that operates on a "last in, first out" principle. All storage locations are initially empty. Two operations, PUSH and POP, describe the subsequent history of the stack. An item of data is added to the top of the stack by PUSH. This pushes previously stored items further down in the stack. Only the topmost item on the stack is accessible, and it is recalled and removed from the stack by POP.

Suppose the sequence $1, 2, \ldots, n$ is run through a stack. Let U and O represent PUSH and POP, respectively. For $n = 3$ the stack sequences and corresponding permutations are

$$\begin{array}{ll}
(U, O, U, O, U, O) & (1, 2, 3) \\
(U, O, U, U, O, O) & (1, 3, 2) \\
(U, U, O, O, U, O) & (2, 1, 3) \\
(U, U, O, U, O, O) & (2, 3, 1) \\
(U, U, U, O, O, O) & (3, 2, 1).
\end{array}$$

The defining property of a stack sequence is that there is no point at which POP has been executed more times than PUSH. Since each operation occurs a total of n times, the number of possible stack sequences is C_n. The stack operations cannot produce a permutation satisfying $i < j < k$ and $\pi(i) > \pi(k) > \pi(j)$. To see this, note that for any three numbers, if the largest comes off the stack first, then at that point the second largest must be on the stack and the third largest beneath it on the stack. However, $\pi(i) > \pi(k) > \pi(j)$ with $i < j < k$ is the only forbidden feature of a stack permutation, so the construction given provides a bijection between the class of restricted permutations of $[n]$ and the ballot sequences of length $2n$. □

3.3 Recurrence Relations

A **recurrence relation** (or, simply, a **recurrence**) for the sequence (a_n) is a relation of the form

$$a_n = f(a_{n-1}, a_{n-2}, \dots, a_{n-p}).$$

Note that the index p might depend on n. The recurrence is either **linear** or **non-linear** according to the form of the function f. The simplest examples are those that are linear and where p is indepedent of n. A further simplification occurs if the coefficients in the linear function are independent of n. In this case, the recurrence relation is

$$a_n = c_1 a_{n-1} + c_2 a_{n-2} + \cdots + c_p a_{n-p},$$

where p and the coefficients $c_1, c_2, \dots, c_p$ are constants. To completely specify the sequence, the initial values $a_0, a_1, \cdots, a_{p-1}$ must be given. The first step is to find a class of special solutions. Substituting the trial solution $a_n = \lambda^n$, we see that a solution is obtained if λ satisfies the polynomial equation

$$\lambda^n = c_1 \lambda^{n-1} + c_2 \lambda^{n-2} + \cdots + c_p \lambda^{n-p},$$

or, since we may assume $\lambda \neq 0$,

$$\lambda^p - c_1 \lambda^{p-1} - c_2 \lambda^{p-2} - \cdots - c_p = 0. \tag{3.1}$$

This is called the **characteristic equation** of the recurrence. In the simplest case, the roots of this equation are distinct numbers $\lambda_1, \lambda_2, \dots, \lambda_p$. In this case, there exist corresponding numbers $r_1, r_2, \dots, r_p$ such that

$$a_n = \sum_{k=1}^{p} r_k \lambda_k^n \qquad n = 0, 1, 2, \dots \tag{3.2}$$

satisfies the recurrence and assumes the given initial values $a_0, a_1, \dots, a_{p-1}$. (To see this last fact, note that by substituting $n = 0, 1, \dots, p-1$ into (3.2) we find that $(r_1, r_2, \dots, r_p)$ gives the solution of a system of p linear equations. The determinant of this system is the **Vandermonde determinant**, which has the value $\prod_{i>j}(\lambda_i - \lambda_j)$. Since we have assumed the λ's are distinct, the determinant is nonzero. Thus the system has a unique solution.) In case

(3.1) has multiple roots, the solution has to be modified. We shall not go into the details.

Example 3.9 *Let*

$$a_n = \sum_{k=0}^{\lfloor n/2 \rfloor} \binom{n-k}{k} x^k, \qquad n \ge 0.$$

Show that for $x \neq -1/4$,

$$a_n = \frac{1}{\sqrt{1+4x}}\left[\left(\frac{1+\sqrt{1+4x}}{2}\right)^{n+1} - \left(\frac{1-\sqrt{1+4x}}{2}\right)^{n+1}\right].$$

Solution. Note that $a_0 = a_1 = 1$. Since $\binom{n-k}{k} = 0$ for $k > \lfloor n/2 \rfloor$, we can write

$$a_n = \sum_{k \ge 0} \binom{n-k}{k} x^k.$$

Using

$$\binom{n-k}{k} = \binom{n-k-1}{k} + \binom{n-k-1}{k-1}$$

we obtain

$$\begin{aligned} a_n &= \sum_{k \ge 0} \binom{n-1-k}{k} x^k + x \sum_{k \ge 1} \binom{n-2-(k-1)}{k-1} x^{k-1} \\ &= a_{n-1} + x a_{n-2}. \end{aligned}$$

The characteristic equation is $\lambda^2 - \lambda - x = 0$, which has distinct roots

$$\lambda_1 = \frac{1+\sqrt{1+4x}}{2}, \qquad \lambda_2 = \frac{1-\sqrt{1+4x}}{2}.$$

The initial conditions yield $r_1 + r_2 = 1$ and $r_1\lambda_1 + r_2\lambda_2 = 1$, from which we obtain

$$r_1 = \frac{\lambda_1}{\sqrt{1+4x}}, \qquad r_2 = -\frac{\lambda_2}{\sqrt{1+4x}}.$$

Thus

$$a_n = r_1\lambda_1^n + r_2\lambda_2^n = \frac{\lambda_1^{n+1} - \lambda_2^{n+1}}{\sqrt{1+4x}} \tag{3.3}$$

is the desired result for $x \neq -1/4$. It is easy to check that for $x = -1/4$ the solution is $a_n = (n+1)2^{-n}$. □

Note. For $x = 1$ the recurrence relation is the one for Fibonacci numbers, so we find

$$F_{n+1} = \sum_{k=0}^{\lfloor n/2 \rfloor} \binom{n-k}{k} \tag{3.4}$$

$$= \frac{1}{\sqrt{5}}\left[\left(\frac{1+\sqrt{5}}{2}\right)^{n+1} - \left(\frac{1-\sqrt{5}}{2}\right)^{n+1}\right]. \tag{3.5}$$

The result

$$F_m = \frac{1}{\sqrt{5}}\left[\left(\frac{1+\sqrt{5}}{2}\right)^{m} - \left(\frac{1-\sqrt{5}}{2}\right)^{m}\right],$$

which appears in an induction exercise in Chapter 1, is known as **Binet's formula**. Note that (3.3) holds for the case where λ_1 and λ_2 are complex. For example, with $x = -1$ the roots are $\lambda_1 = e^{i\pi/3}$ and $\lambda_2 = e^{-i\pi/3}$, and (3.3) yields

$$a_n = \frac{2\sin((n+1)\pi/3)}{\sqrt{3}} = \begin{cases} 1 & n \equiv 0, 1 \pmod 6 \\ 0 & n \equiv 2, 5 \pmod 6 \\ -1 & n \equiv 3, 4 \pmod 6). \end{cases}$$

Example 3.10 (Terquem's Problem) *Find the number of subsets of $[n]$ having the property that, when the elements are put in increasing order, the smallest one is odd, the next smallest is even, and so on. (Sets with this property are said to be* **alternating**. *By convention, the empty set is alternating.)*

Solution. Let a_n denote the number of alternating subsets of $[n]$. The first few values are $a_0 = 1$, $a_1 = 2$, $a_2 = 3$, $a_3 = 5$, $a_4 = 8$. Comparison with the Fibonacci sequence leads to the conjecture that $a_n = F_{n+2}$. Clearly a_{n-1} alternating subsets of $[n]$ do not contain the element n. We claim that a_{n-2} alternating subsets of $[n]$ do contain n. To see this, consider the mapping from the alternating subsets of $[n]$ containing n to the alternating subsets of $[n-2]$ obtained by simply deleting n and deleting $n-1$ as well if it is an element of the given set. In the other direction, given an alternating subset of $[n-2]$, add

n if the largest element of the given subset has parity opposite to that of n and add both $n-1$ and n if the parity of the largest element is the same as n. It is easy to see that the given mapping is a bijection, so the number of alternating subsets of $[n]$ containing n is a_{n-2}. Thus (a_n) satisfies the recurrence

$$a_n = a_{n-1} + a_{n-2};$$

since $a_0 = 1$ and $a_1 = 2$, the desired solution is $a_n = F_{n+2}$. □

Let π be a permutation of $[n]$. Since π a bijection, at some point the sequence of iterates $\pi(1)$, $\pi(\pi(1))$, $\pi(\pi(\pi(1)))$, ... returns to 1 and completes a **cycle**. If we pick an element not on this cycle (provided there is one) and do the same thing, the iterates generate a second disjoint cycle. Continuing in this way, we find that a given permutation is thus composed of disjoint cycles. A cycle with k elements will be called a k-cycle. A 1-cycle is called a **fixed point**. There are many interesting combinatorial problems concerning cycles in permutations. One of the most famous has to do with counting permutations with no fixed points.

Example 3.11 (Derangements) *Let D_n denote the number of permutations of $[n]$ with no fixed points. Show that*

$$D_n = n! \sum_{k=0}^{n} \frac{(-1)^k}{k!}.$$

Solution. The first few values are $D_1 = 0$, $D_2 = 1$, $D_3 = 2$, $D_4 = 9$. We claim that

$$D_{n+1} = n(D_n + D_{n-1}), \qquad n \geq 2.$$

To see this, first note that if π is a permutation of $[n+1]$ with no fixed points then $\pi(n+1)$ has one of n possible values. Suppose $\pi(n+1) = i$. Then there are two cases: (i) $\pi(i) = n+1$ so that $n+1$ is on a 2-cycle, (ii) $n+1$ is on a k-cycle where $k > 2$. In the first case, there are clearly D_{n-2} corresponding permutations. In the second case, there is some $j \neq i$ such that $\pi(j) = n+1$. We obtain a corresponding permutation of $[n]$ by modifying the given permutation to read $\pi(j) = i$. The resulting correspondence is a bijection from the set of all permutations of $[n+1]$ with no fixed points and satisfying $\pi(n+1) = i \neq n+1$ onto the set of permutations of $[n]$ with no fixed points.

Since there are n choices for i, the recurrence is established. From this recurrence, we find

$$D_{n+1} - (n+1)D_n = -(D_n - nD_{n-1}),$$

with $D_2 - 2D_1 = 1$, so by induction $D_n - nD_{n-1} = (-1)^n$. Dividing both sides by $n!$ and summing, we obtain

$$\frac{D_m}{m!} = \sum_{n=2}^{m} \left(\frac{D_n}{n!} - \frac{D_{n-1}}{(n-1)!} \right) = \sum_{n=2}^{m} \frac{(-1)^n}{n!} = \sum_{n=0}^{m} \frac{(-1)^n}{n!}. \quad \square$$

Here is an example of a linear recurrence with nonconstant coefficients.

Example 3.12 *Let a_n denote the number of permutations π of $[n]$ such that $\pi(\pi(i)) = i$ for $i = 1, 2, \ldots, n$. Find a recurrence for (a_n).*

Solution. A permutation satisfying $\pi(\pi(i)) = i$ for all i is called an **involution**. Suppose that $\pi : [n] \to [n]$ is an involution. If $\pi(n) = n$ then the restriction of π to $[n-1]$ is also an involution; there are a_{n-1} such permutations. If $\pi(n) = i \neq n$ then $\pi(i) = n$ and the restriction of π to $\{1, 2, \ldots, i-1, i+1, \ldots, n-1\}$ is an involution. Since there are $n-1$ choices for i, there are $(n-1)a_{n-2}$ permutations π of $[n]$ that are involutions satisfying $\pi(n) \neq n$. Thus the desired recurrence is

$$a_n = a_{n-1} + (n-1)a_{n-2}. \quad \square$$

Using the method of generating functions (§3.4) and a little bit of calculus, one can use this recurrence to show that

$$\sum_{n=0}^{\infty} \frac{a_n}{n!} x^n = e^{x+x^2/2}.$$

Also, we can find an explicit formula for a_n. In an involution, each cycle is either a fixed point or a 2-cycle. If there are k 2-cycles, there are $\binom{n}{2k}$ ways to choose the points on the 2-cycles and $(2k)!/(k!2^k)$ ways to pair them up. (Here we use the overcounting technique illustrated in Example 3.1.) It follows that

$$a_n = n! \sum_{k=0}^{\lfloor n/2 \rfloor} \frac{1}{2^k k!(n-2k)!}.$$

An important example of a nonlinear recurrence is that satisfied by the Catalan sequence. We illustrate the derivation of this recurrence with a problem involving the triangularization of convex polygons. A set K in the plane is **convex** if for any two points $x, y \in K$ the entire line segment joining x to y is also in K. The boundary of a convex n-gon consists of n straight line segments, called the **sides** of the polygon that meet at **vertices** of the polygon. A **diagonal** is a line segment joining two nonconsecutive vertices.

Example 3.13 *Find the number of ways to divide a convex $(n+2)$-gon into triangles by diagonals that do not intersect in the interior of the polygon.*

Solution. Let a_n denote the number of ways to divide a convex $(n+2)$-gon into triangles as specified. By convention, we take $a_0 = a_1 = 1$. The next few values are $a_2 = 2$, $a_3 = 5$, $a_4 = 14$. We thus suspect that the numbers in question are the Catalan numbers. Given a convex $(n+2)$-gon, number its vertices $0, 1, \ldots, n+1$ consecutively. Consider the side joining vertices n and $n+1$. If the triangle containing this side has vertex k as its remaining point, then the diagonals joining vertices n and $n+1$ to vertex k define, in addition to the given triangle, a convex $(k+2)$-gon and convex $(n+1-k)$-gon. There are thus $a_k a_{n-1-k}$ ways to finish the division of the original $(n+2)$-gon into triangles. Summing on k, we have

$$a_n = \sum_{k=0}^{n-1} a_k \, a_{n-1-k}.$$

The solution of this recurrence is

$$a_n = \frac{1}{n+1}\binom{2n}{n} = C_n.$$

We shall prove this using generating functions in §3.4. □

The following solution uses several techniques, including recurrence and the multisection formula.

Example 3.14 (Proposed for the 1992 IMO) *For $k \geq 1$ let a_k be the greatest divisor of k that is not a multiple of 3. Set $S_0 = 0$ and $S_k = a_1 + a_2 + \cdots + a_k$ for $k \geq 1$. Let A_n denote the number of integers $0 \leq k < 3^n$ such that S_k is divisible by 3. Show that*

$$A_n = \frac{3^n + 2 \cdot 3^{n/2} \cos(n\pi/6)}{3}.$$

Solution. Note that $a_{3k} = a_k$ whereas $a_{3k+1} = 3k + 1$ and $a_{3k+2} = 3k + 2$. Also $S_1 \equiv 1 \pmod 3$ and $S_2 \equiv 0 \pmod 3$. We claim that for each $k \geq 0$,

$$S_{3k} \equiv S_k, \qquad S_{3k+1} \equiv S_k + 1, \qquad S_{3k+2} \equiv S_k \pmod 3.$$

This is true for $k = 0$, and for $k \geq 1$ mathematical induction yields

$$\begin{aligned} S_{3k} &= S_{3k-1} + a_{3k} \equiv S_{k-1} + a_k \equiv S_k \pmod 3, \\ S_{3k+1} &= S_{3k} + a_{3k+1} \equiv S_k + 1 \pmod 3, \\ S_{3k+2} &= S_{3k+1} + a_{3k+2} \equiv (S_k + 1) + 2 \equiv S_k \pmod 3. \end{aligned}$$

Let T_k denote the number of ones in the ternary (base 3) representation of k. Then $T_0 = 0$, $T_1 = 1$, and $T_2 = 0$. Also, a moment's thought shows that

$$T_{3k} = T_k, \qquad T_{3k+1} = T_k + 1, \qquad T_{3k+2} = T_k.$$

It follows that $S_k \equiv T_k \pmod 3$, so A_n is the number of integers $0 \leq k < 3^n$ that have a ternary representation in which the number of ones is a multiple of 3. There are $\binom{n}{j}2^{n-j}$ nonnegative integers less than 3^n that have j ones in their ternary representation since there are $\binom{n}{j}$ ways to choose the positions for the ones and then 2^{n-j} ways to choose the digits (zeros or twos) to fill the remaining places. It follows that

$$A_n = \sum_{j \equiv 0 \pmod 3} \binom{n}{j} 2^{n-j}.$$

To evaluate A_n we use the multisection formula (§1.5). Let

$$f(x) = (x + 2)^n = \sum_{j=0}^{n} \binom{n}{j} x^k 2^{n-j}$$

and $\omega = (-1 + i\sqrt{3})/2$. The multisection formula yields

$$\begin{aligned} A_n &= \sum_{j \equiv 0 \pmod 3} \binom{n}{j} 2^{n-j} \\ &= \frac{f(1) + f(\omega) + f(\overline{\omega})}{3} \\ &= \frac{3^n + (\sqrt{3}e^{i\pi/6})^n + (\sqrt{3}e^{-i\pi/6})^n}{3} \end{aligned}$$

$$= \frac{3^n + 2 \cdot 3^{n/2}\cos(n\pi/6)}{3}. \quad \square$$

3.4 Generating Functions

3.4.1 Generating Functions from Recurrence Relations

Let $a_0, a_1, a_2, \ldots$ be a sequence of interest. We write

$$G(x) = \sum_{n=0}^{\infty} a_n x^n,$$

and call this series the **ordinary generating function** for the given sequence. The accepted theoretical framework for the use of generating functions is that of **formal power series**. In this view, the method involves algebraic relations satisfied by the sequence $(a_0, a_1, a_2, \ldots)$ and not analytic properties of the generating function, so convergence is not an issue.

To symbolize that a_n is the coefficient of x^n in the given generating function, we write

$$a_n = [x^n]G(x).$$

For example, the generating function for the Fibonacci sequence is

$$G(x) = \sum_{n=1}^{\infty} F_n x^n = \frac{x}{1 - x - x^2}. \tag{3.6}$$

To prove this, we use the recurrence relation for Fibonacci numbers to find

$$\begin{aligned} G(x) &= x + x^2 + \sum_{n=3}^{\infty}(F_{n-1} + F_{n-2})x^n \\ &= x + x^2 + x(G(x) - x) + x^2 G(x). \end{aligned}$$

Solving for $G(x)$ we get (3.6). By partial fractions,

$$\frac{x}{1 - x - x^2} = \frac{1}{\sqrt{5}}\left[\frac{1}{1 - ax} - \frac{1}{1 - bx}\right] = \frac{1}{\sqrt{5}}\sum_{n=1}^{\infty}(a^n - b^n)x^n,$$

where

$$a = \frac{1+\sqrt{5}}{2}, \qquad b = \frac{1-\sqrt{5}}{2}.$$

Thus we have

$$F_n = [x^n]\frac{x}{1-x-x^2} = \frac{a^n - b^n}{\sqrt{5}}$$

(Binet's formula) as before.

As this example suggests, generating functions provide another general method for solving recurrence relations.

Example 3.15 (Proposed for the 1992 IMO) *Let a_n be the number of words of length n consisting of symbols 0 and 1 such that neither 101 nor 111 occur as 3-digit blocks. Express a_n in terms of Fibonacci numbers.*

Solution. By convention, we take $a_0 = 1$. The next few terms are $a_1 = 2$, $a_2 = 4$, $a_3 = 6$, $a_4 = 9$. If a word of length n begins with 0, the remaining word of length $n-1$ can be any one fulfilling our condition; thus there are a_{n-1} such words. If the word begins with 10, the next symbol must be 0 and the remaining word of length $n-3$ can be any one of the a_{n-3} allowed words. If the word begins with 11, the next two symbols must be 0 and the remaining word of length $n-4$ can be any one of a_{n-4} possibilities. Thus we obtain the recurrence relation

$$a_n = a_{n-1} + a_{n-3} + a_{n-4}, \qquad n \geq 4.$$

It follows that the generating function $G(x) = \sum_{n=0}^{\infty} a_n x^n$ satisfies

$$\begin{aligned} G(x) &= 1 + 2x + 4x^2 + 6x^3 + \sum_{n=4}^{\infty}(a_{n-1} + a_{n-3} + a_{n-4})x^n \\ &= 1 + 2x + 4x^2 + 6x^3 + x(G(x) - 1 - 2x - 4x^2) \\ &\quad + x^3(G(x) - 1) + x^4 G(x), \end{aligned}$$

and from this we obtain

$$G(x) = \frac{1 + x + 2x^2 + x^3}{1 - x - x^3 - x^4} = \frac{1 + x + 2x^2 + x^3}{(1 - x - x^2)(1 + x^2)}.$$

By partial fractions,

$$G(x) = \frac{1}{5}\left[\frac{7 + 4x}{1 - x - x^2} - \frac{2 + x}{1 + x^2}\right].$$

Now we can just read off the answer. We have

$$[x^n]\frac{7+4x}{1-x-x^2} = 7F_{n+1} + 4F_n = F_{n+5} + F_{n+3}$$

(by repeated use of the Fibonacci recurrence) and

$$[x^n]\frac{2+x}{1+x^2} = \begin{cases} 2(-1)^{n/2}, & n \text{ even,} \\ (-1)^{(n-1)/2}, & n \text{ odd,} \end{cases}$$

so

$$a_n = \begin{cases} \dfrac{F_{n+5}+F_{n+3}-2(-1)^{n/2}}{5} & \text{if } n \text{ is even,} \\ \dfrac{F_{n+5}+F_{n+3}-(-1)^{(n-1)/2}}{5} & \text{if } n \text{ is odd.} \end{cases} \qquad \square$$

Note. There are various equivalent ways to write this result. Specifically, the solution can be expressed by the formulas $a_{2k} = F_{k+2}^2$ and $a_{2k-1} = F_{k+1}F_{k+2}$. It is possible to conjecture these formulas on the basis of small cases and then complete the proof by showing that they satisfy the recurrence. Thus

$$\begin{aligned} a_{2k} &= a_{2k-1} + a_{2k-3} + a_{2k-4} = F_{k+1}F_{k+2} + F_kF_{k+1} + F_k^2 \\ &= F_{k+1}F_{k+2} + F_kF_{k+2} = F_{k+2}^2, \end{aligned}$$

and

$$\begin{aligned} a_{2k-1} &= a_{2k-2} + a_{2k-4} + a_{2k-5} = F_{k+1}^2 + F_k^2 + F_{k-1}F_k \\ &= F_{k+1}^2 + F_kF_{k+1} = F_{k+1}F_{k+2}. \end{aligned}$$

The advantage of the generating function method is its straightforwardness. No conjectures are required; just calculate the generating function from the recurrence and then extract its coefficients by partial fractions.

The process of obtaining generating functions from recurrence relations is illustrated by the study of various combinatorial numbers. The number of partitions of $[n]$ into k nonempty blocks is the **Stirling number of the second kind** $\left\{ {n \atop k} \right\}$. First of all we need a recurrence relation, namely

$$\left\{ {n \atop k} \right\} = \left\{ {n-1 \atop k-1} \right\} + k\left\{ {n-1 \atop k} \right\}. \qquad (3.7)$$

To see this, consider the placement of the element n in a partition of $[n]$ into k blocks. The first term on the right counts those partitions where n is in a block by itself, so the remaining $n-1$ elements are in $k-1$ blocks. The second term counts the partitions in which n is in a block along with at least one other element; for each partition of $[n-1]$ into k blocks, we can choose one of the k blocks into which n is inserted. This recurrence relation holds for all $n \geq 1$ if we take

$$\left\{ {0 \atop 0} \right\} = 1, \qquad \left\{ {0 \atop k} \right\} = 0 \text{ for } k \geq 1.$$

Let

$$G_k(x) = \sum_{n=0}^{\infty} \left\{ {n \atop k} \right\} x^n.$$

Then $G_0(x) = 1$, and multiplying both sides of (3.7) by x^n and summing, we find

$$G_k(x) = G_{k-1}(x) + kxG_k(x).$$

Thus $(1-kx)G_k(x) = G_{k-1}(x)$, and it follows by induction that

$$G_k(x) = \frac{1}{(1-x)(1-2x)\cdots(1-kx)}.$$

The number of permutations of $[n]$ with k cycles is the (unsigned) **Stirling number of the first kind** $\left[{n \atop k} \right]$. Here the appropriate recurrence is

$$\left[{n \atop k} \right] = \left[{n-1 \atop k-1} \right] + (n-1)\left[{n-1 \atop k} \right]. \tag{3.8}$$

To see this, consider the status of the element n in a permutation of $[n]$ with k cycles. The first term on the right counts those permutations where n is in a cycle by itself. The second term counts those permutations where n belongs to a cycle with other elements. Given such a permutation, by deleting n and closing up the cycle we get a permutation of $[n-1]$ with k cycles. In the other direction, there are $n-1$ choices for where n might be inserted. Let

$$H_n(x) = \sum_{k=1}^{n} \left[{n \atop k} \right] x^k.$$

Then $H_1(x) = x$, and mutliplying both sides of (3.8) by x^k and summing, we obtain

$$H_n(x) = (x+n-1)H_{n-1}(x).$$

Thus by induction

$$\sum_{k=1}^{n} \begin{bmatrix} n \\ k \end{bmatrix} x^k = H_n(x) = x(x+1)\cdots(x+n-1). \qquad (3.9)$$

The right hand side is called a **rising factorial** and is denoted by $x^{\overline{n}}$.

Example 3.16 *Find the value of the sum*

$$\sum_{S\in[n]^k} \frac{1}{\pi(S)}$$

where $\pi(S)$ denotes the product of the elements of S and the sum is over all k-element subsets of $[n]$. By convention $\pi(\varnothing) = 1$.

Solution. Consider the polynomial

$$P_n(x) = \left(1+\frac{x}{1}\right)\left(1+\frac{x}{2}\right)\cdots\left(1+\frac{x}{n}\right).$$

A typical term in the product is

$$\frac{x^k}{a_1 a_2 \cdots a_k}$$

where $\{a_1, a_2, \ldots, a_k\} \subseteq [n]$. It follows that

$$\sum_{S\in[n]^k} \frac{1}{\pi(S)} = [x^k]P_n(x) = \frac{1}{n!}[x^{k+1}]H_{n+1}(x) = \frac{1}{n!}\begin{bmatrix} n+1 \\ k+1 \end{bmatrix}. \quad \square$$

Stirling numbers of the first kind provide the answer to another interesting problem about permutations. Let π be a permutation of $[n]$. A **left-to-right maximum** of π occurs whenever $\pi(j) < \pi(i)$ for all $j < i$.

Example 3.17 *Prove that the number of permutations of $[n]$ with k left-to-right maxima is* $\begin{bmatrix} n \\ k \end{bmatrix}$.

Solution. We give two proofs, the first of which uses the concept of an **inversion table**. Given a permutation π, an **inversion** in π is a pair (i, j) such that $i < j$ and $\pi(i) > \pi(j)$. For $k = 1, 2, \ldots, n$

let a_k be the number of inversions (i, j) where $\pi(j) = k$. Then the elements in the inversion table satisfy $0 \le a_k \le n - k$. There is a simple bijection between the $n!$ permutations of $[n]$ and the $n!$ possible inversion tables. To see this, we have to see how to recover π uniquely from its inversion table. Just insert the the integers $n, n-1, \ldots, 1$ in their proper order according to the inversion table. For example, if $n = 6$ and $(3, 4, 3, 2, 1, 0)$ is the inversion table, we can build up the permutation one step at a time as follows: (6), $(6, 5)$, $(6, 5, 4)$, $(6, 5, 4, 3)$, $(6, 5, 4, 3, 2)$, $(6, 5, 4, 1, 3, 2)$. Note that the number of left-to-right maxima in π is the same as the number of zeros in the corresponding inversion table.

Let x be a positive integer and let B_x denote the set of all $(x + 1)(x + 2) \cdots (x + n)$ sequences $(b_1, b_2, \ldots, b_n)$ satisfying

$$0 \le b_k \le n + x - k, \qquad k = 1, 2, \ldots, n.$$

Consider the mapping that associates with each element of B_x a corresponding sequence $(a_1, a_2, \ldots, a_n)$ given by

$$a_k = \max(0, b_k - x), \qquad k = 1, 2, \ldots, n.$$

Note that $(a_1, a_2, \ldots, a_n)$ satisfies $0 \le a_k \le n - k$, $1 \le k \le n$, and thus it is the inversion table of some corresponding permutation π. If $a_k = 0$ then b_k can be any one of the values $0, 1, \ldots, x$ and if $a_k > 0$ then b_k is determined by $b_k = a_k + x$. It follows that if an inversion table has exactly L zeros then its inverse image in the given mapping contains $(x + 1)^L$ elements of B_x. Since the number of zeros in the inversion table is the same as the number of left-to-right maxima in the corresponding permutation, we have

$$\sum_{\pi} (x + 1)^{L(\pi)} = (x + 1)(x + 2) \cdots (x + n).$$

Both sides are polynomials of degree n in x and thus (replacing $x + 1$ by x)

$$\sum_{\pi} x^{L(\pi)} = x(x + 1) \cdots (x + n - 1)$$

is an identity. Comparing this with (3.9) we obtain the desired result. □

Every permutation can be written as the product of disjoint cycles. To standardize the representation, begin each cycle with its largest element and put these largest elements in increasing order. Thus, for example, the permutation

$$\pi = \begin{pmatrix} 1 & 2 & 3 & 4 & 5 & 6 \\ 3 & 5 & 4 & 1 & 6 & 2 \end{pmatrix}$$

has cycles $(1, 3, 4)$ and $(2, 5, 6)$. The standard form is $(4, 1, 3)(6, 2, 5)$. Erasing the parentheses, we obtain 413625, and thus the corresponding permutation

$$\hat{\pi} = \begin{pmatrix} 1 & 2 & 3 & 4 & 5 & 6 \\ 4 & 1 & 3 & 6 & 2 & 5 \end{pmatrix}$$

having two left-to-right maxima.

This procedure works in general. Given a permutation, (i) write the cycle representation with the largest element of each cycle first and these largest elements in increasing order, (ii) remove the parentheses. The resulting mapping is a bijection that associates with each permutation having k cycles another permutation with k left-to-right maxima. This completes the second proof. □

3.4.2 Direct Calculation of Generating Functions

Instead of going through the step of setting up a recurrence relation, it is sometimes possible to make a direct connection between the generating function and the set of combinatorial objects it describes. Let S be a set and let w be a function from S to the set of nonnegative integers. We shall refer to $w(\sigma)$ as the **weight** of $\sigma \in S$. The generating function for S with respect to w is defined by

$$G_S(x) = \sum_{\sigma \in S} x^{w(\sigma)}.$$

For example, if S is the set of permutations of $\{1, 2, 3\}$ and $w(\sigma)$ is the number of fixed points of σ, then the elements and their weights are as shown in the following table.

Element	Weight
(1,2,3)	3
(1,3,2)	1
(2,1,3)	1
(2,3,1)	0
(3,1,2)	0
(3,2,1)	1

Thus the generating function is

$$G_S(x) = x^3 + x + x + 1 + 1 + x = x^3 + 3x + 2.$$

The basic Sum and Product Rules have counterparts in the land of generating functions.

Rule 3.3 (Sum) *If $S = A \cup B$ where $A \cap B = \emptyset$ then*

$$G_S(x) = G_A(x) + G_B(x).$$

Rule 3.4 (Product) *If $S = A \times B$ and if for all $\sigma \in S$ the weight of $\sigma = (a, b)$ is $w(a) + w(b)$ then*

$$G_S(x) = G_A(x)G_B(x).$$

The Product Rule is based on the following algebraic fact: if

$$G(x) = \sum_{n=0}^{\infty} a_n x^n \quad \text{and} \quad H(x) = \sum_{n=0}^{\infty} b_n x^n,$$

then

$$[x^n]G(x)H(x) = \sum_{k=0}^{n} a_k b_{n-k}.$$

As before, these generalize to $S = A_1 \cup A_2 \cup \cdots \cup A_k$ and $S = A_1 \times A_2 \times \cdots A_k$, respectively. In preparation for the next example, we record a very important series expansion, namely the general binomial expansion

$$(1 - x)^{-a} = \sum_{n=0}^{\infty} \binom{n + a - 1}{n} x^n, \qquad |x| < 1.$$

Historical Note. This result was discovered by Isaac Newton in the winter of 1664 and communicated to Henry Oldenberg, the secretary of the Royal Society, in 1677. A general proof was given by Neils Henrik Abel 150 years later.

Example 3.18 *Let S be the set of all k-tuples $(a_1, a_2, \ldots, a_k)$ where the a_i's are positive integers and the weight of the element $\sigma = (a_1, a_2, \ldots, a_k)$ in S is $\sum_{i=1}^{n} a_i$. Find the generating function.*

Solution. Let $A = \mathbb{Z}^+$. If $k = 1$, the generating function is clearly

$$G_A(x) = x + x^2 + x^3 + \cdots = \frac{x}{1-x}.$$

For arbitrary $k \in \mathbb{Z}^+$ we have $S = A \times A \times \cdots \times A$ with k factors, so by the product rule

$$G_S(x) = \left(\frac{x}{1-x}\right)^k.$$

Since the k-tuple $(a_1, a_2, \ldots, a_k)$ has weight $\sum_{i=1}^{k} a_i$, the coefficient of x^n in $G_S(x)$ is the number of ways to write n as the sum of k positive integers (where the order of the summands counts). A representation of this sort is called a **composition**. Now we can use Newton's series to find the number of compositions of n into positive integers, namely

$$[x^n]G_S(x) = [x^n]\left(\frac{x}{1-x}\right)^k = [x^{n-k}](1-x)^{-k} = \binom{n-1}{k-1}.$$

If the k-tuple consists of nonnegative integers, the number of compositions is

$$[x^n]\left(\frac{1}{1-x}\right)^k = \binom{n+k-1}{k-1}. \quad \square$$

Example 3.19 (Proposed for the 1986 IMO) *Let $S(r, n, k)$ denote the number of k-tuples of nonnegative integers $(x_1, x_2, \ldots, x_k)$ satisfying $x_1 + x_2 + \cdots + x_k = n$ and $x_i \le r$, $i = 1, 2, \ldots, k$. Show that*

$$S(k, n, r) = \sum_{j=0}^{m} (-1)^j \binom{k}{j} \binom{k + n - (r+1)j - 1}{k-1},$$

where $m = \min(k, \lfloor n/(r+1) \rfloor)$.

Solution. By the same considerations as in the last example, the generating function is

$$G_S(x) = (1 + x + \cdots + x^r)^k = \left(\frac{1 - x^{r+1}}{1 - x}\right)^k.$$

It is only necessary to evaluate $[x^n]G_S(x)$. To this end, we expand

$$(1 - x^{r+1})^k = \sum_{j=0}^{k} (-1)^j \binom{k}{j} x^{(r+1)j}$$

and

$$(1 - x)^{-k} = \sum_{s=0}^{\infty} \binom{k + s - 1}{k - 1} x^s,$$

and so obtain

$$\begin{aligned} S(k, n, r) &= [x^n] \left(\sum_{j=0}^{k} (-1)^j \binom{k}{j} x^{(r+1)j} \right) \left(\sum_{s=0}^{\infty} \binom{k + s - 1}{k - 1} x^s \right) \\ &= \sum_{j=0}^{m} (-1)^j \binom{k}{j} \binom{k + n - (r + 1)j - 1}{k - 1}. \end{aligned}$$

(For each j, a contribution is made to the coefficient of x^n when $s + (r + 1)j = n$.) Note that $j \le k$ and $(r + 1)j \le n$; thus $j \le \min(k, \lfloor n/(r + 1) \rfloor)$. □

A **partition** of a positive integer n is a nonincreasing sequence of positive integers whose sum is n. Thus the partition

$$\lambda = (\lambda_1, \lambda_2, \ldots, \lambda_M)$$

satisfies $n = \lambda_1 + \lambda_2 + \cdots + \lambda_M$ where the λ_i's are positive integers and $\lambda_1 \ge \lambda_2 \ge \cdots \ge \lambda_M$. The number of partitions of n is denoted by $p(n)$. It is convenient to define $p(0) = 1$. Then the first few values are $p(0) = 1$, $p(1) = 1$, $p(2) = 2$, $p(3) = 3$, $p(4) = 5$, $p(5) = 7$, $p(6) = 11$. The function $p(n)$ grows rather rapidly. Godfrey H. Hardy and Srinivasa Ramanujan proved the asymptotic formula

$$p(n) \sim \frac{1}{4n\sqrt{3}} e^{\pi\sqrt{2n/3}} \qquad (n \to \infty).$$

To obtain the generating function for partitions, let us first consider partitions where each summand, or **part**, is at most k. The partition is then described by a k-tuple $(a_1, a_2, \ldots, a_k)$ where a_1 is the number of ones in the partition, a_2 is the number of twos, and so on. Thus the a_i's are nonnegative integers satisfying $a_1 + 2a_2 + \cdots + ka_k = n$. Let $L(\lambda)$ denote the largest part (λ_1) of the partition λ and let $p(n; L \leq k)$ denote the number of partitions of n such that each part is at most k. By the Sum and Product Rules, the generating function for partitions where each part is at most k is

$$\sum_{n=0}^{\infty} p(n; L \leq k)x^n = \left(\sum_{a_1=0}^{\infty} x^{a_1}\right)\left(\sum_{a_2=0}^{\infty} x^{2a_2}\right)\cdots\left(\sum_{a_k=0}^{\infty} x^{ka_k}\right)$$

$$= \frac{1}{(1-x)(1-x^2)\cdots(1-x^k)}.$$

If $k \geq n$, then $p(n; L \leq k) = p(n)$. Thus the generating function for partitions is

$$\sum_{n=0}^{\infty} p(n)x^n = \frac{1}{(1-x)(1-x^2)(1-x^3)\cdots}.$$

This result was found by Leonhard Euler, who founded the generating function approach to partition problems. Euler used the generating function method to solve various partition problems in his book *Introductio in Analysin Infinitorum* (1748).

Now that we have worked through the process of finding the generating function for all partitions, it is easy to write down generating functions for various restricted classes of partitions. For example, the generating function for partitions into odd parts is

$$P_{\mathcal{O}}(x) = \frac{1}{(1-x)(1-x^3)(1-x^5)\cdots},$$

and the generating function for partitions into distinct parts is

$$P_{\mathcal{D}}(x) = (1+x)(1+x^2)(1+x^3)\cdots$$

Writing the latter as

$$P_{\mathcal{D}}(x) = \frac{1-x^2}{1-x} \cdot \frac{1-x^4}{1-x^2} \cdot \frac{1-x^6}{1-x^3} \cdots$$

and canceling common factors, we get $P_{\mathcal{D}}(x) = P_{\mathcal{O}}(x)$, and thus find as Euler did that the number of partitions of n into odd parts equals the number of partitions of n into distinct parts.

James Sylvester found a combinatorial proof that there are just as many partitions of n into odd parts as there are partitions of n into distinct parts.

Example 3.20 (Sylvester) *Give a bijective proof that the number of partitions of n into odd parts equals the number of partitions of n into distinct parts.*

Solution. A partition in which every part is odd is described by the sequence $a_1, a_3, a_5, \ldots$ where a_{2k+1} is the number of parts equal to $2k + 1$. Given such a partition, expand each of the numbers $a_1, a_3, a_5, \ldots$ in binary and constuct a corresponding partition in which $(2p + 1)2^q$ is a part whenever 2^q is a term in the binary expansion of a_{2p+1}. Since $(2p + 1)2^q = (2r + 1)2^s$ implies $p = r$ and $q = s$, this construction gives a partition of n into distinct parts. In the other direction, given a partition of n into distinct parts, write each part as the product of an odd integer and a power of two, and let the result $(2p + 1)2^q$ signify that 2^q is a term in the binary expansion of a_{2p+1}. □

Besides the method of generating functions, there is another widely applicable technique for proving results about partitions. This approach involves a graphical representation of partitions due to Norman Macleod Ferrers. To represent the partition $n = \lambda_1 + \lambda_2 + \cdots + \lambda_M$ where $\lambda_1 \geq \lambda_2 \geq \cdots \geq \lambda_M$, construct a diagram in which there are λ_1 dots in the first row, λ_2 dots in the second row, and so on, with the diagram left-justified. For example, the partition $25 = 7 + 7 + 5 + 3 + 2 + 1$ is represented by the **Ferrers diagram** shown below.

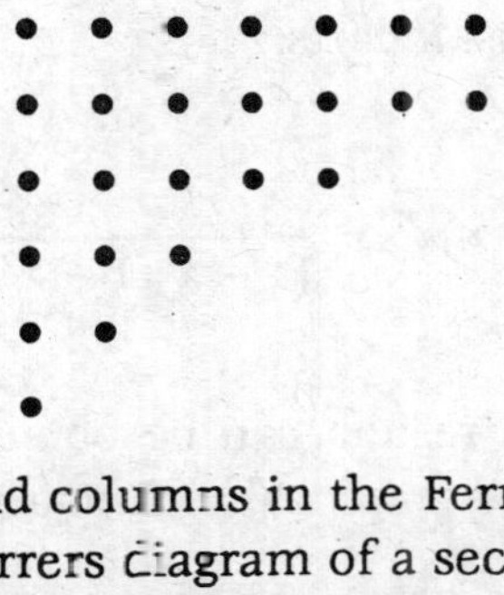

$$
\begin{array}{ccccccc}
\bullet & \bullet & \bullet & \bullet & \bullet & \bullet & \bullet \\
\bullet & \bullet & \bullet & \bullet & \bullet & \bullet & \bullet \\
\bullet & \bullet & \bullet & \bullet & \bullet & & \\
\bullet & \bullet & \bullet & & & & \\
\bullet & \bullet & & & & & \\
\bullet & & & & & &
\end{array}
$$

Interchanging rows and columns in the Ferrers diagram of a given partition yields the Ferrers diagram of a second partition, which is said to be **conjugate** to the first one. For example, interchanging rows and columns in the Ferrer diagram above gives the one shown below, corresponding to the partition 25 = 6 + 5 + 4 + 3 + 3 + 2 + 2.

```
• • • • • •
• • • • •
• • • •
• • •
• • •
• •
• •
```

Denote the conjugate of $\lambda = (\lambda_1, \lambda_2, \ldots, \lambda_p)$ by $\lambda' = (\lambda_1', \lambda_2', \ldots, \lambda_q')$. A partition λ that satisfies $\lambda = \lambda'$ is said to be **self-conjugate**. Conjugation maps a partition whose largest part is at most k to one with at most k parts. Thus we have the appropriate bijection to show that these two classes of partitions are equinumerous. Let $M(\lambda)$ denote the number of parts of the partition λ and let $p(n; M \leq k)$ denote the the number of partitions of n into at most k parts. From the previously obtained generating function, we have

$$\sum_{n=0}^{\infty} p(n; M \leq k)x^n = \frac{1}{(1-x)(1-x^2)\cdots(1-x^k)}.$$

The generating function for partitions into exactly k parts then follows by subtracting the generating function for at most $k - 1$ parts from the one for at most k parts:

$$\sum_{n=0}^{\infty} p(n; M = k)x^n = \frac{x^k}{(1-x)(1-x^2)\cdots(1-x^k)}.$$

Some very elegant results can be obtained by combining the generating function and graphical techniques.

Example 3.21 *Find the generating function for self-conjugate partitions.*

Solution. As an example of a self-conjugate partition, consider $17 = 6 + 4 + 3 + 2 + 1 + 1$, whose Ferrers diagram is shown below.

```
• • • • • •
• • • •
• • •
• •
•
•
```

The 3×3 square of dots in the upper left corner is called the **Durfee square** of this partition. In general, the Durfee square of the partition $(\lambda_1, \lambda_2, \ldots, \lambda_p)$ is the $k \times k$ square in the upper left corner of the Ferrers diagram where $k = \max\{i \mid \lambda_i \geq i\}$. The deletion of the Durfee square leaves the Ferrers diagrams of two partitions, one into at most k parts and the other with largest part at most k. If the original partition is self-conjugate, these two are conjugate partitions of $(n - k^2)/2$. (In the example above, the deletion of the Durfee square leaves the Ferrers diagrams of $4 = 3 + 1$ and $4 = 2 + 1 + 1$.) Thus the Ferrers diagram of a self-conjugate partition is completely specified by giving the size of the Durfee square (k) and the identity of a partition (μ) of $(n-k^2)/2$ into at most k parts. It follows that the number of self-conjugate partitions of n for which the Durfee square has size k is $p((n-k^2)/2; M \leq k)$. With this in mind, we are ready to write down the generating function for self-conjugate partitions:

$$G_{SC}(x) = 1 + \sum_{k=1}^{\infty} \frac{x^{k^2}}{(1-x^2)(1-x^4)\cdots(1-x^{2k})}. \quad \square$$

Note that to count all partitions (not just the self-conjugate ones) using this same Durfee square idea, we can use the fact that

$$\left(\frac{1}{(1-x)(1-x^2)\cdots(1-x^k)}\right)^2$$

is the generating function for ordered pairs of partitions, each with at most k parts. Thus we obtain the interesting identity

$$\frac{1}{(1-x)(1-x^2)(1-x^3)\cdots} = 1 + \sum_{k=1}^{\infty} \frac{x^{k^2}}{((1-x)(1-x^2)\cdots(1-x^k))^2}.$$

The method described in this section provides an efficient method for solving problems involving 0–1 strings. To see how this works, let's first introduce some notation. Let ϵ denote the empty string. If a and b are 0–1 strings, then ab is the string obtained by concatenating a and b. For example, if $a = 101$ and $b = 1011$ then $ab = 1011011$. If A and B are sets of strings, then $AB = \{ab \mid a \in A,\ b \in B\}$. We shall let A^* denote the set of all strings obtainted by concatenating any number of strings from A:

$$A^* = \epsilon \cup A \cup AA \cup AAA \cup \cdots$$

Consider generating functions for sets of 0–1 strings where the weight of a string is its length. If $G_A(x)$ and $G_B(x)$ are two such generating functions, then it is not true in general that $G_{AB}(x) = G_A(x)G_B(x)$. However, if each element of AB has a *unique* representation of the form ab, where $a \in A$ and $b \in B$, then AB can be identified with $A \times B$, and thus we have

$$G_{AB}(x) = G_A(x)G_B(x).$$

For example, if $A = \{0, 00, 000, \ldots\}$ and $B = \{1, 111, 11111, \ldots\}$ then

$$G_A(x) = \frac{x}{1-x}, \qquad G_B(x) = \frac{x}{1-x^2},$$

and

$$G_{AB}(x) = G_A(x)G_B(x) = \frac{x^2}{(1-x)(1-x^2)}.$$

It is easily checked that this generating function correctly enumerates 0–1 strings consisting of a nonempty block of zeros followed by a block consisting of an odd number of ones. Also, for certain A's, the set A^* is uniquely generated, so

$$G_{A^*}(x) = 1 + G_A(x) + (G_A(x))^2 + \cdots = \frac{1}{1 - G_A(x)}.$$

For example, if $A = 11^*00^*$ then $G_A(x) = (x/(1-x))^2$ and $G_{A^*}(x) = 1/(1-G_A(x)) = (1-x)^2/(1-2x)$. Since any 0–1 string can be thought of as consisting of alternating blocks of zeros and ones, we can construct generating functions for various classes of such strings by using the Sum Rule and this version of the Product Rule. Let's apply this idea to the set of all 0–1 strings. We may assume that the string begins with a possibly empty block of zeros (0^*) and that it ends with a possibly empty block of ones (1^*). What happens in between? Well, maybe nothing (ϵ). Otherwise, we have some alternating nonempty blocks of ones and zeros. Thus the set of all 0–1 strings is symbolized by

$$0^*[\epsilon \cup 11^*00^* \cup 11^*00^* \cup \cdots]1^* = 0^*[11^*00^*]^*1^*.$$

On the basis of this representation, we can use the Sum Rule and Product Rule to write down the corresponding generating function:

$$G(x) = \frac{1}{1-x}\left[\frac{1}{1-(x/(1-x))^2}\right]\frac{1}{1-x} = \frac{1}{1-2x}.$$

At this point the reader is no doubt asking how the above procedure could be described as "efficient." After all, it is completely obvious that there are 2^n 0–1 strings of length n, so the generating function has to be $\sum_{n=0}^{\infty} 2^n x^n = 1/(1-2x)$. The answer to this objection is that now that the groundwork has been laid, we have an efficient method for finding the generating functions to count restricted 0–1 strings. We can use the symbolism developed to describe the appropriate class of 0–1 strings and then write down the generating function by inspection.

Example 3.22 *Find the generating function for the set of all 0–1 strings with no two consecutive zeros.*

Solution. The set of strings in question is represented by $(\epsilon \cup 0)[11^*0]^*1^*$, and thus the generating function is

$$G(x) = (1+x)\left[\frac{1}{1-(x^2/(1-x))}\right]\frac{1}{1-x} = \frac{1+x}{1-x-x^2}.$$

It follows that the number of strings of length n is $[x^n]G(x) = F_{n+1} + F_n = F_{n+2}$. □

Now we give a second solution of the problem considered in Example 3.4.

Example 3.23 *Find the number of 0–1 strings of length n with exactly m zeros that are followed immediately by ones.*

Solution. This set of strings is represented by 1*[00*11*]^{m}0*, so the generating function is

$$\frac{1}{1-x}\left(\frac{x^2}{(1-x)^2}\right)^m\frac{1}{1-x}=\frac{x^{2m}}{(1-x)^{2m+2}}.$$

Thus the number of such strings of length n is

$$[x^n]\frac{x^{2m}}{(1-x)^{2m+2}}=[x^{n-2m}](1-x)^{-(2m+2)}=\binom{n+1}{2m+1}. \quad \square$$

Using this approach, let's reconsider the problem solved in Example 3.15.

Example 3.24 *Find the generating function for the set of 0–1 strings where neither the patterns 101 nor 111 occur.*

Solution. The desired representation is

$$0^*\cup 0^*(1\cup 11)[000^*(1\cup 11)]^*0^*,$$

from which we obtain the generating function

$$\begin{aligned}G(x)&=\frac{1}{1-x}+\frac{x+x^2}{1-x}\left[1-\frac{x^2(x+x^2)}{1-x}\right]^{-1}\frac{1}{1-x}\\&=\frac{1+x+2x^2+x^3}{1-x-x^3-x^4},\end{aligned}$$

as before. $\square$

The usefulness of generating functions is not limited to counting problems. In the following problem, $A\Delta B$ denotes the **symmetric difference** of sets A and B defined by $A\Delta B=(A\cup B)-(A\cap B)$.

Example 3.25 *Let $A_1, A_2, \ldots$ and $B_1, B_2, \ldots$ be sets of integers defined as follows: $A_1=\varnothing$, $B_1=\{0\}$ and*

$$\begin{aligned}A_{n+1}&=\{k+1\,|\,k\in B_n\},\\B_{n+1}&=A_n\Delta B_n,\end{aligned}$$

for $n\geq 1$. Find all n such that $B_n=\{0\}$.

Solution. We shall prove that $B_n=\{0\}$ if and only if n is a power of two. Introduce the generating functions

$$F_n(x)=\sum_{k\in A_n}x^k,\qquad G_n(x)=\sum_{k\in B_n}x^k,\qquad n\geq 1.$$

The recursive definitions of the sequences (A_n) and (B_n) translate into simple recurrence formulas for the generating functions (mod 2):

$$F_{n+1}(x) \equiv xG_n(x), \qquad G_{n+1}(x) \equiv F_n(x) + G_n(x) \pmod 2.$$

Thus $G_1(x) = G_2(x) = 1$ and

$$G_{n+2} \equiv G_{n+1}(x) + xG_n(x) \pmod 2$$

for $n \geq 1$. Using

$$\binom{n-k+1}{k} = \binom{n-k}{k} + \binom{n-k}{k-1},$$

it is easy to show by induction that

$$G_n(x) \equiv \sum_k \binom{n-k-1}{k} x^k \pmod 2,$$

so

$$B_n = \left\{ k \,\middle|\, \binom{n-k-1}{k} \equiv 1 \pmod 2 \right\}.$$

Thus it suffices to prove that $\binom{n-k-1}{k}$ is even for every $k > 0$ if and only if n is a power of two. At this point we could use the theorem about parity of binomial coefficients proved in §1.1 (and again in §1.3). Here is an alternative proof. We use Newton's expansion

$$(1+x)^{-a} = \sum_{k=0}^{\infty} \binom{k+a-1}{k} (-1)^k x^k,$$

and the fact that $(1+x)^{2^r} \equiv 1 + x^{2^r} \pmod 2$. If n is a power of two and $0 < k < n/2$ then

$$\begin{aligned}
\binom{n-k-1}{k} &\equiv [x^k](1+x)^{n-k-1} \\
&\equiv [x^k]\frac{1+x^n}{(1+x)^{k+1}} \\
&\equiv [x^k](1+x)^{-(k+1)} \\
&\equiv \binom{2k}{k} \\
&\equiv 0 \pmod 2.
\end{aligned}$$

The last step follows from $\binom{2k}{k} = \binom{2k-1}{k-1} + \binom{2k-1}{k} = 2\binom{2k-1}{k}$. Of course if $k \geq n/2$ then $\binom{n-k-1}{k} = 0$. Thus $B_n = \{0\}$ if n is a power of two. If n is not a power of two, we can write $n = 2^r + s$ where $0 < s < 2^r$. Then

$$\begin{aligned}\binom{n-s-1}{s} &\equiv [x^s](1+x)^{2^r-1} \\ &\equiv [x^s]\frac{1+x^{2^r}}{1+x} \\ &\equiv [x^s](1+x)^{-1} \\ &\equiv 1 \pmod 2,\end{aligned}$$

so $s \in B_n$ and $B_n \neq \{0\}$. □

Finally, to fulfill a promise made in Example 3.13, we use generating functions to show that if $a_0 = 1$ and

$$a_n = \sum_{k=0}^{n-1} a_k a_{n-1-k}, \qquad n \geq 1, \tag{3.10}$$

then

$$a_n = \frac{1}{n+1}\binom{2n}{n}.$$

Let $G(x) = \sum_{n=0}^{\infty} a_n x^n$. We multiply both sides of (3.10) by x^n and sum:

$$G(x) = 1 + x\sum_{n=1}^{\infty}\left\{\sum_{k=0}^{n-1} a_k a_{n-1-k}\right\}x^{n-1} = 1 + xG^2(x).$$

There are two roots of this equation, but it is perfectly clear which one we want:

$$G(x) = \frac{1-\sqrt{1-4x}}{2x}. \tag{3.11}$$

To obtain $a_n = [x^n]G(x)$, we first note that $\binom{n+1/2}{n}4^n = \binom{2n}{n}$ so Newton's expansion for $(1-4x)^{-1/2}$ can be written as

$$(1-4x)^{-1/2} = \sum_{n=0}^{\infty}\binom{2n}{n}x^n.$$

Thus from (3.11) we have

$$\begin{aligned} a_n &= -\frac{1}{2}[x^{n+1}]\sqrt{1-4x} \\ &= -\frac{1}{2}[x^{n+1}](1-4x)\sum_{n=0}^{\infty}\binom{2n}{n}x^n \\ &= \frac{1}{2}\left[4\binom{2n}{n} - \binom{2n+2}{n+1}\right] \\ &= \left[2 - \frac{2n+1}{n+1}\right]\binom{2n}{n} \\ &= \frac{1}{n+1}\binom{2n}{n}. \end{aligned}$$

There are many occasions where the generating function method can be teamed with the multisection formula to good advantage. A good example is provided by a problem on the 1995 IMO.

Example 3.26 (1995 IMO) *Let p be an odd prime number. Find the number of subsets A of the set $\{1, 2, 3, \ldots, 2p\}$ such that (i) A has exactly p elements, and (ii) the sum of all the elements in A is divisible by p.*

Solution. We shall give two solutions of a modest generalization of the problem. Let $\sigma(A)$ denote the sum of all the elements in A. In the generalized problem we ask for the number of subsets A of $[mp]$ such that $|A| = p$ and $\sigma(A) \equiv 0 \pmod p$. Let

$$C_m = \{A \subseteq [mp] \mid |A| = p,\ \sigma(A) \equiv 0 \pmod p\}.$$

Using the multisection formula, we shall show that

$$|C_m| = \frac{1}{p}\left(\binom{mp}{p} + m(p-1)\right).$$

Consider the polynomial

$$F(x,y) = (1+xy)(1+xy^2)\cdots(1+xy^{mp}).$$

Each contribution to $x^k y^\sigma$ in the product corresponds to a subset (possibly empty) $A = \{i_1, i_2, \ldots, i_k\} \subseteq [mp]$ with $|A| = k$ and $\sigma(A) = \sigma$. It follows that

$$|C_m| = \sum_{\sigma \equiv 0 \pmod p} [x^p y^\sigma]F(x,y).$$

The rest is algebra. By the multisection formula,

$$|\mathcal{C}_m| = [x^p]\frac{1}{p}\sum_{j=0}^{p-1} F(x, \omega^j),$$

where $\omega = e^{2\pi i/p}$. Thus

$$\begin{aligned}|\mathcal{C}_m| &= [x^p]\frac{1}{p}\sum_{j=0}^{p-1}(1 + x\omega^j)(1 + x\omega^{2j})\cdots(1 + x\omega^{mpj})\\ &= [x^p]\frac{1}{p}\sum_{j=0}^{p-1}\left\{(1 + x)(1 + x\omega^j)\cdots(1 + x\omega^{(p-1)j})\right\}^m,\end{aligned}$$

where the last step uses the fact that $\omega^{kj} = \omega^{(k+p)j}$. For $j = 0$ the summand is $(1 + x)^{mp}$, and for each of the remaining $p-1$ values of j the summand is $(1 + x^p)^m$. To see this, note that $(1, \omega^j, \omega^{2j}, \ldots, \omega^{(p-1)j})$ is a permutation of $(1, \omega, \omega^2, \ldots, \omega^{p-1})$ and

$$(1 + x)(1 + \omega x)(1 + \omega^2 x)\cdots(1 + \omega^{p-1}x) = 1 + x^p.$$

Hence

$$\begin{aligned}|\mathcal{C}_m| &= [x^p]\frac{1}{p}\left\{(1 + x)^{mp} + (p - 1)(1 + x^p)^m\right\}\\ &= \frac{1}{p}\left(\binom{mp}{p} + m(p - 1)\right).\end{aligned}$$

Setting $m = 2$, we obtain the answer to the original question:

$$|\mathcal{C}_2| = \frac{1}{p}\left(\binom{2p}{p} + 2(p - 1)\right). \quad \square$$

Note. One USA student, Jacob Lurie, used the multisection formula to find a formula that holds when the prime p is replaced by an arbitrary positive integer n. This formula involves the Euler phi function.

The second solution uses more conventional combinatorial methods. Let p be an odd prime and for $0 \le k \le p$ and $0 \le r \le p-1$ let

$$\mathcal{F}_{k,r} = \{A \subseteq [p] \mid |A| = k,\ \sigma(A) \equiv r \pmod{p}\}.$$

First we prove that for $1 \le k \le p-1$,

$$|\mathcal{F}_{k,0}| = |\mathcal{F}_{k,1}| = \cdots = |\mathcal{F}_{k,p-1}| = \frac{1}{p}\binom{p}{k}.$$

Since $\sum_{r=0}^{p-1} |\mathcal{F}_{k,r}| = \binom{p}{k}$, it suffices to show that for $0 \le r, s \le p-1$ there is a bijection $\phi : \mathcal{F}_{k,r} \to \mathcal{F}_{k,s}$. For $A \in \mathcal{F}_{k,r}$ let

$$\phi(A) = \{x + k^{-1}(s-r) | \ x \in A\}$$

where the calculation of $x + k^{-1}(s-r)$ is performed in $\mathbb{Z}_p$. Clearly $\sigma(A) \equiv r \pmod p$ gives $\sigma(\phi(A)) \equiv s \pmod p$ so $\phi : \mathcal{F}_{k,r} \to \mathcal{F}_{k,s}$, and it is easy to check that ϕ is a bijection. The inverse mapping is $\phi^{-1}(B) = \{y + k^{-1}(r-s) | \ y \in B\}$.

Let

$$C_m = \{A \subseteq [mp] | \ |A| = p, \ \sigma(A) \equiv 0 \pmod p\}$$

as before. We shall find a formula for $|C_m|$ by establishing a recurrence relation. Assume $m > 1$. For $A \in C_m$ let $X = A \cap [p]$ and $Y = A - X$. We divide the contributions to $|C_m|$ into three cases. (i) In the first case, $X = A = [p]$. This contributes 1 to the count of all sets A belonging to C_m. (ii) In the second case, $X = \varnothing$. Then $A \subseteq \{p+1, p+2, \ldots, mp\}$ and it is easy to see that the number of choices for A in this case is $|C_{m-1}|$. (iii) Finally, if $|X| = k$ where $1 \le k \le p-1$ there are $\binom{(m-1)p}{p-k}$ choices for Y, and whatever choice is made, the residue class of $\sigma(X)$ modulo p is then determined by the fact that

$$\sigma(A) = \sigma(X) + \sigma(Y) \equiv 0 \pmod p.$$

By the previous result there are $\binom{p}{k}/p$ choices for X. Putting these observations together, we have

$$|C_m| = 1 + |C_{m-1}| + \frac{1}{p}\sum_{k=1}^{p-1}\binom{(m-1)p}{p-k}\binom{p}{k}.$$

To evaluate the sum we use **Vandermonde's identity**

$$\sum_{k=0}^{n}\binom{a}{k}\binom{b}{n-k} = \binom{a+b}{n}.$$

The proof of this result is given as Exercise 1(a), §3.2-3.5. We thus obtain

$$\sum_{k=1}^{p-1}\binom{(m-1)p}{p-k}\binom{p}{k}=\binom{mp}{p}-\binom{(m-1)p}{p}-1.$$

Consequently,

$$|\mathcal{C}_m|-|\mathcal{C}_{m-1}|=\frac{1}{p}\left(\binom{mp}{p}-\binom{(m-1)p}{p}+(p-1)\right).$$

Trivially, $|\mathcal{C}_1| = 1$, so the above relation holds for $m = 1$ if we take $\mathcal{C}_0 = \varnothing$. Thus by telescoping sums,

$$|\mathcal{C}_m|=\frac{1}{p}\left(\binom{mp}{p}+m(p-1)\right)$$

Substitution of $m = 2$ yields the solution of the original problem. □

3.5 The Inclusion-Exclusion Principle

Suppose $P_1, P_2, \ldots, P_n$ are subsets of S and we would like to know how many elements of S belong to precisely r of these subsets. (The subsets $P_1, P_2, \ldots, P_n$ may be identified with corresponding properties of the elements of S; thus P_i consists of those elements having property P_i.) Often it is easy to count the number of elements in arbitrary intersections of the P_i's. This informaiion can be used to answer the above question. Specifically, for $I \subseteq [n]$, let

$$M_k=\sum_{|I|=k}M(\supseteq I)\qquad\text{where}\qquad M(\supseteq I)=\left|\bigcap_{i\in I}P_i\right|.$$

Then the number of elments of S that belong to precisely r of the P_i's is given by

$$E_r=\sum_{k=r}^{n}(-1)^{k-r}\binom{k}{r}M_k. \tag{3.12}$$

This is the **inclusion-exclusion formula**. To prove it we first note that

$$M_k = \sum_{r=k}^{n} \binom{r}{k} E_r$$

since an element belonging to exactly r of the P_i's contributes $\binom{r}{k}$ times to the sum defining M_k. We introduce the polynomials

$$M(x) = \sum_{k=0}^{n} M_k x^k \quad \text{and} \quad E(x) = \sum_{r=0}^{n} E_r x^r.$$

Then

$$\begin{aligned} M(x) &= \sum_{k=0}^{n} \sum_{r=k}^{n} \binom{r}{k} E_r x^k \\ &= \sum_{r=0}^{n} E_r \sum_{k=0}^{r} \binom{r}{k} x^k \\ &= \sum_{r=0}^{n} E_r (x+1)^r \\ &= E(x+1). \end{aligned}$$

It follows that

$$\begin{aligned} E(x) &= M(x-1) \\ &= \sum_{k=0}^{n} M_k (x-1)^k \\ &= \sum_{k=0}^{n} M_k \sum_{r=0}^{k} \binom{k}{r} x^r (-1)^{k-r} \\ &= \sum_{r=0}^{n} x^r \sum_{k=r}^{n} (-1)^{k-r} \binom{k}{r} M_k, \end{aligned}$$

from which we obtain (3.12).

One of the standard applications of the inclusion-exclusion formula yields the result for derangement numbers found in §3.3. There are $\binom{n}{k}$ choices for $I \subseteq [n]$ with $|I| = k$, and for each one there are $(n-k)!$ permutations of $[n]$ such that $\pi(i) = i$ for all $i \in I$. Thus $M_k = \binom{n}{k}(n-k)! = n!/k!$ and we obtain for the nth derangement number

$$D_n = \sum_{k=0}^{n}(-1)^k M_k = n! \sum_{k=0}^{n} \frac{(-1)^k}{k!},$$

as before. More generally, the number of permutations of $[n]$ having exactly r fixed points is given by

$$E_r = \sum_{k=r}^{n}(-1)^{k-r}\binom{k}{r}\frac{n!}{k!} = \frac{n!}{r!}\sum_{j=0}^{n-r}\frac{(-1)^j}{j!} = \binom{n}{r}D_{n-r}.$$

This result can be obtained directly by observing that for a permutation of $[n]$ with exactly r fixed points there are $\binom{n}{r}$ choices for the fixed points and then D_{n-r} choices for the (fixed-point free) permutation of the remaining $n-r$ points.

Example 3.27 *Show that the number of permutations π of $[n]$ such that $\pi(j+1) \neq \pi(j)+1$ for $j = 1, 2, \ldots, n-1$ is $D_n + D_{n-1}$.*

Solution. For $i = 1, 2, \ldots, n-1$ say that π has property P_i if $\pi(j) = i$ and $\pi(j+1) = i+1$ for some j. For any choice of k of these properties, the number of permutations of $[n]$ that have these k properties (and possibly more) is $(n-k)!$. To see this, think of the blocks of consecutive integers required by the k properties as glued together and permuted as a unit. Thus, for example, permutations of [10] that have properties P_1, P_2, P_5, P_6, P_8 are obtained by permuting the five elements (1, 2, 3), 4, (5, 6, 7), (8, 9), 10. It follows by the inclusion-exclusion formula that the number of permutations of $[n]$ that have none of the properties $P_1, P_2, \ldots, P_{n-1}$ is

$$\begin{aligned}\sum_{k=0}^{n-1}(-1)^k\binom{n-1}{k}(n-k)! &= (n-1)!\sum_{k=0}^{n}(n-k)\frac{(-1)^k}{k!} \\ &= n!\sum_{k=0}^{n}\frac{(-1)^k}{k} + (n-1)!\sum_{k=1}^{n}\frac{(-1)^{k-1}}{(k-1)!} \\ &= D_n + D_{n-1}. \quad \square\end{aligned}$$

Another application of the inclusion-exclusion formula gives the following expression for Stirling numbers of the second kind.

Example 3.28 *Prove that*

$$\left\{ n \atop k \right\} = \frac{1}{k!} \sum_{j=0}^{k} (-1)^j \binom{k}{j} (k-j)^n. \tag{3.13}$$

Solution. We first use the inclusion-exclusion formula to count the number of arrangements of the elements of $[n]$ into k labeled boxes so that no box is empty. Consider the set of all arrangements, whether or not any box is empty, and for $1 \le i \le k$ let P_i be the subset where box i is empty. If $|I| = j$, there are $\binom{k}{j}$ choices for $I \subseteq [k]$, and for any such choice the number of arrangements where box i is empty for all $i \in I$ is $(k-j)^n$ since each element of $[n]$ then occupies one of $k-j$ boxes. Dividing by $k!$ accounts for the fact that the boxes are unlabeled, and thus we have (3.13). □

There are other ways to obtain this result. One is by inverting the following relation satisfied by Stirling numbers of the second kind:

$$x^n = \sum_k \left\{ n \atop k \right\} x(x-1)(x-2)\cdots(x-k+1). \tag{3.14}$$

The product $x(x-1)(x-2)\cdots(x-k+1)$ is called a **falling factorial** and is denoted by $x^{\underline{k}}$. To prove (3.14), let n and r be positive integers and consider the set of all r^n functions from $[n]$ into $[r]$. Given $f : [n] \to [r]$, let

$$\mathcal{R}(f) = \{y \in [r] | \; y = f(x) \text{ for some } x \in [n]\}$$

denote its range. Each function f with range of size k corresponds to an arrangement of the elements of $[n]$ into k labeled boxes. There are $k! \left\{ n \atop k \right\}$ such arrangements and $\binom{n}{k}$ ways to choose the elements belonging to $\mathcal{R}(f)$. There are thus

$$\left\{ n \atop k \right\} \binom{r}{k} k! = \left\{ n \atop k \right\} r^{\underline{k}}$$

functions $f : [n] \to [r]$ with $|\mathcal{R}(f)| = k$. Summing over k, we have

$$r^n = \sum_{k=0}^{n} \left\{ n \atop k \right\} r^{\underline{k}}. \tag{3.15}$$

(By including $k = 0$ in this sum, the result holds true for all $n \ge 0$. For $n \ge 1$ this term makes no contribution.) This formula holds for arbitrary positive integers r, and each side is a polynomial of degree n. Thus

$$x^n = \sum_{k=0}^{n} \left\{ {n \atop k} \right\} x^{\underline{k}} = \sum_k \left\{ {n \atop k} \right\} \binom{x}{k} k!$$

holds for arbitrary $x \in \mathbb{C}$. To give a second proof of (3.13), we first note the binomial coefficient identity

$$\begin{aligned} \sum_{j=l}^{k} (-1)^{k-j} \binom{k}{j} \binom{j}{l} &= (-1)^{k-l} \binom{k}{l} \sum_{j=l}^{k} \binom{k-l}{j-l} (-1)^{j-l} \\ &= \begin{cases} 1, & k = l, \\ 0, & k \neq l. \end{cases} \end{aligned}$$

Thus

$$\begin{aligned} \sum_{j=0}^{k} (-1)^{k-j} \binom{k}{j} j^n &= \sum_{j=0}^{k} (-1)^{k-j} \binom{k}{j} \sum_{l=0}^{n} \left\{ {n \atop l} \right\} \binom{j}{l} l! \\ &= \sum_{l=0}^{n} l! \left\{ {n \atop l} \right\} \sum_{j=l}^{k} (-1)^{k-j} \binom{k}{j} \binom{j}{l} \\ &= k! \left\{ {n \atop k} \right\}, \end{aligned}$$

which again yields (3.13). Note that (3.9) gives

$$\sum_{k=1}^{n} (-1)^{n-k} \left[{n \atop k} \right] x^k = x^{\underline{n}}.$$

Comparing this formula with (3.14) and using the linear independence of the polynomials $x^{\underline{k}}$, $k = 0, 1, \ldots, n$, we find

$$\sum_{k=m}^{n} (-1)^{n-k} \left[{n \atop k} \right] \left\{ {k \atop m} \right\} = \begin{cases} 1, & m = n, \\ 0, & m \neq n. \end{cases}$$

Thus these two kinds of combinatorial numbers are related by more than the fact that they both are named after James Stirling.

Example 3.29 *How many $m \times n$ 0–1 matrices have no row or column in which every element is 0?*

Solution. Consider $m \times n$ matrices in which no row consists entirely of zeros. There are $2^n - 1$ possibilities for each row, so there are altogether $(2^n - 1)^m$ such matrices. For any k columns, the number of $m \times n$ 0–1 matrices in which those columns (and possibly more) consist entirely of zeros is $(2^{n-k} - 1)^m$. It follows that

$$M_k = \binom{n}{k}(2^{n-k} - 1)^m$$

and

$$E_0 = \sum_{k=0}^{n}(-1)^k\binom{n}{k}(2^{n-k} - 1)^m.$$

Note that by expanding $(2^{n-k} - 1)^m$ the result can be written as

$$E_0 = \sum_{k=0}^{n}\sum_{j=0}^{m}(-1)^{k+j}\binom{n}{k}\binom{m}{j}2^{(m-j)(n-k)}. \tag{3.16}$$

This is the form obtained by applying inclusion-exclusion to the set of all $m \times n$ 0–1 matrices. For $|J| = j$ and $|K| = k$ the number of matrices where all rows with indices in J and all columns with indices in K consist entirely of zeros is $M(\supseteq (J, K)) = 2^{(m-j)(n-k)}$. The basic inclusion-exclusion formula then yields the double sum (3.16). □

The proof of the inclusion-exclusion formula was based on the the relation $E(x) = M(x - 1)$ where

$$E(x) = \sum_{r=0}^{n} E_r x^r \qquad \text{and} \qquad M(x) = \sum_{k=0}^{n} M_k x^k.$$

At the time, the polynomials $E(x)$ and $M(x)$ were just means to an end. However, note that if $\sigma \in S$ is given the weight $w(\sigma) = |\{i \mid \sigma \in P_i\}|$ then

$$E(x) = \sum_{\sigma \in S} x^{w(\sigma)}$$

is the generating function for S with respect to w. Thus there is a natural connection between generating functions and the inclusion-exclusion formula. As an example, let's reconsider the problem at the beginning of §3.4.2.

Example 3.30 *Find the generating function*

$$G_n(x) = \sum_{\sigma \in S_n} x^{w(\sigma)},$$

where S_n denotes the set of all permutations of $[n]$ and $w(\sigma)$ is the number of fixed points of σ.

Solution. Using

$$M_k = \binom{n}{k}(n-k)! = \frac{n!}{k!},$$

we have

$$G_n(x) = n! \sum_{k=0}^{n} \frac{(x-1)^k}{k!}. \quad \square$$

The following problem from the 1994 USAMO provides a good opportunity to use the inclusion-exclusion formula.

Example 3.31 (1994 USAMO) *Let $|U|$, $\sigma(U)$, and $\pi(U)$ denote the number of elements, the sum, and the product, respectively, of a finite set of positive integers. (If U is the empty set, $|U| = 0$, $\sigma(U) = 0$, $\pi(U) = 1$.) Let S be a finite set of positive integers. Prove that*

$$\sum_{U \subseteq S} (-1)^{|U|} \binom{m - \sigma(U)}{|S|} = \pi(S)$$

for all integers $m \geq \sigma(S)$.

Solution. Let $S = \{a_1, a_2, \ldots, a_n\}$. For 0–1 sequences of length $m \geq \sigma(S)$, let the first a_1 positions in the sequence be block 1, the next a_2 positions be block 2, and so on. How many 0–1 sequences have n ones and $m - n$ zeros with each of the n blocks accounting for a single 1? There are clearly $a_1 a_2 \cdots a_n = \pi(S)$ such sequences. Now let's count the same set of sequences using the inclusion-exclusion formula. For $I \subseteq [n]$ set $\sigma(I) = \sum_{i \in I} a_i$ and $U(I) = \{a_i | i \in I\}$. The number of 0–1 sequences of length m where there are n ones and, for each $i \in I$, block i consists entirely of zeros, is

$$M(\supseteq I) = \binom{m - \sigma(I)}{n} = \binom{m - \sigma(U)}{|S|}$$

since the n ones can be chosen to be in any of $m - \sigma(I)$ positions. It follows from the inclusion-exclusion formula that

$$\sum_{I \subseteq [n]} (-1)^{|I|} \binom{m - \sigma(I)}{n} = a_1 a_2 \cdots a_n = \pi(S). \quad \square \tag{3.17}$$

Note. With the interpretation

$$\binom{x}{n} = \frac{x(x-1)\cdots(x-n+1)}{n!},$$

(3.17) is an algebraic identity and holds not only for the case where $a_1, a_2,$
$\ldots, a_n$ and m are positive integers with $m \geq a_1 + a_2 + \cdots + a_n$ but also for arbitrary real or complex values of these parameters.

Here is a generating function solution of the same problem. Let $n = |S|$. By the Binomial Theorem,

$$\sum_{U \subseteq S} (-1)^{|U|} \binom{m - \sigma(U)}{|S|} = [x^n] \sum_{U \subseteq S} (-1)^{|U|} (1 + x)^{m - \sigma(U)}.$$

Thus we have reduced the problem to working out the coefficient of $[x^n]$ in the given generating function. Since the product $\prod_{a \in S} \{(1+x)^a - 1\}$ expands to yield $\pi(S)x^n$ plus higher order terms, we have

$$\begin{aligned}
\sum_{U \subseteq S} (-1)^{|U|} \binom{m - \sigma(U)}{|S|} &= [x^n] \sum_{U \subseteq S} (-1)^{|U|} (1 + x)^{m - \sigma(U)} \\
&= [x^n](1 + x)^m \sum_{U \subseteq S} \frac{(-1)^{|U|}}{(1 + x)^{\sigma(U)}} \\
&= [x^n](1 + x)^m \prod_{a \in S} \left\{1 - \frac{1}{(1 + x)^a}\right\} \\
&= [x^n](1 + x)^{m - \sigma(S)} \prod_{a \in S} \{(1 + x)^a - 1\} \\
&= [x^n](1 + x)^{m - \sigma(S)} \prod_{a \in S} (ax + \cdots) \\
&= [x^n](1 + x)^{m - \sigma(S)} \left(\pi(S)x^n + \cdots\right) \\
&= \pi(S). \quad \square
\end{aligned}$$

Exercises for Sections 3.2-3.5

1. Prove the following facts involving binomial coefficients.
   (a) $\sum_k \binom{m}{k}\binom{n}{r-k} = \binom{m+n}{r}$.
   (b) $\sum_k \binom{m}{k}\binom{n+k}{m} = \sum_j \binom{m}{j}\binom{n}{j}2^j$.
   The first result is known as **Vandermonde's identity**.
2. (a) How many ballot sequences with n A's and n B's are there where A and B are never tied until the last vote? (b) How many such sequences are there where A and B are tied at exactly one point before the last vote?
3. A random walk on $\mathbb{Z} \times \mathbb{Z}$ starts at $(0, 0)$ and reaches (a, b) after $n = a + b + 2c$ steps, never having left the first quadrant ($x, y \geq 0$). Show that the number of walks is
$$W_n(a, b) = \frac{n!(a + 1)(b + 1)}{(a + c + 1)!(b + c + 1)!}\binom{n + 2}{c}.$$
   Hint: First show that the number of acceptable walks with $h = a + 2r$ horizontal steps and $v = b + 2s$ vertical steps is
$$\binom{n}{h}\frac{a + 1}{a + r + 1}\binom{h}{r}\frac{b + 1}{b + s + 1}\binom{v}{s},$$
   which simplifies to
$$\frac{n!(a + 1)(b + 1)}{(a + c + 1)!(b + c + 1)!}\binom{b + c + 1}{r}\binom{a + c + 1}{s}.$$
   Now use Vandermonde's identity.
4. How many strings of length n can be formed with the alphabet $\{0, 1, 2, 3, 4\}$ if neighboring digits differ by 1 (in absolute value)? [Proposed for the 1987 IMO]
5. For each $n \geq 1$, find the sum of the products $F_{k_1}F_{k_2}\cdots F_{k_r}$ where the sum is over all 2^{n-1} compositions $n = k_1 + k_2 + \cdots + k_r$. (For example, for $n = 3$ the desired sum is $F_3 + F_1 \cdot F_2 + F_2 \cdot F_1 + F_1 \cdot F_1 \cdot F_1 = 5$.)
6. Find the number of 0–1 strings of length n containing no block consisting of an odd number of zeros between two nonempty blocks of ones.

7. Call a finite set of positive integers S **fat** if every element of S is at least $|S|$. By convention, the empty set is fat. Since a k-element fat subset of $[n]$ must be chosen from $\{k, k+1, \ldots, n\}$, by (3.4) the total number of fat subsets of $[n]$ is

$$\sum_{k=0}^{\lfloor (n+1)/2 \rfloor} \binom{n-k+1}{k} = F_{n+2}.$$

There are also F_{n+2} alternating subsets of $[n]$ (see Example 3.10). How many subsets of $[n]$ are both alternating and fat?

8. Given $k \geq 1$, let a_n be the number of 0–1 strings of length n that do not have k consecutive zeros, and let b_n be the number of 0–1 strings that have neither $k+1$ consecutive zeros nor $k+1$ consecutive ones. Prove that $b_{n+1} = 2a_n$.

9. Show that the number of subsets of $[n]$ containing exactly one pair of consecutive integers is

$$\sum_{k=1}^{n-1} F_k F_{n-k} = \frac{2nF_{n+1} - (n+1)F_n}{5}.$$

10. Find the sequence (a_n) if $a_0 = 1$ and

$$\sum_{k=0}^{n} a_k a_{n-k} = 1, \qquad n \geq 1.$$

11. Prove that the number of partitions of n into parts not divisible by d is the same as the number of partitions of n in which no part occurs d or more times. [Glaisher]

12. Prove that the number of partitions of n in which all the even parts are distinct is the same as the number of partitions of n in which each part is repeated at most three times. [Andrews]

13. Prove that the number of partitions of n into distinct parts all of which are odd is the same as the number of partitions of n that are self-conjugate. [Sylvester]

14. Find the number of permutations of $[n]$ that have no r-cycle.

15. Using the inclusion-exclusion formula or otherwise, prove

$$\sum_{k=r}^{n} (-1)^{k-r} \binom{k}{r} \binom{n-k}{k} 2^{n-2k} = \binom{n+1}{2r+1}.$$

3.6 The Pigeonhole Principle

The Pigeonhole Principle is sometimes called **Dirichlet's Principle** since Lejeune Dirichlet is credited with first realizing that this simple principle could be used to establish nontrivial results.

Principle 3.2 (Pigeonhole Principle) *If n objects are placed in k different boxes, then at least one of the boxes contains $\lceil n/k \rceil$ or more objects.*

We begin by discussing a problem to which Dirichlet applied this principle with notable success. This takes us away from combinatorics for a moment, but it is a worthwhile digression.

Theorem 3.1 (Dirichlet) *Let α be an irrational number. Then there are infinitely many integer pairs (h, k) where $k > 0$ such that*

$$\left|\alpha - \frac{h}{k}\right| < \frac{1}{k^2}. \tag{3.18}$$

Proof. Let Q be an arbitrary positive integer. We claim that there exist integers h and k, with $1 \leq k \leq Q$, such that

$$\left|\alpha - \frac{h}{k}\right| < \frac{1}{kQ}. \tag{3.19}$$

This implies that the desired infinite collection of rational numbers exists. Otherwise, there would be a positive number ϵ such that $|\alpha - h/k| > \epsilon$ for all (h, k) satisfying (3.18). But then, by choosing $Q > 1/\epsilon$, (3.19) yields

$$\left|\alpha - \frac{h}{k}\right| < \frac{1}{kQ} \leq \min\left(\frac{1}{k^2}, \frac{1}{Q}\right) \leq \min\left(\frac{1}{k^2}, \epsilon\right),$$

a contradiction. Now we need to prove the claim, and here is where the Pigeonhole Principle is used. The "boxes" are the intervals

$$B_k = \left\{x \,\middle|\, \frac{k-1}{Q} \leq x < \frac{k}{Q}\right\}, \qquad k = 1, 2, \ldots, Q,$$

and the "objects" are the numbers $\{q\alpha\}$ ($q = 0, 1, 2, \ldots, Q$), where $\{x\} = x - \lfloor x \rfloor$ denotes the fractional part of x. Since there are Q boxes and $Q + 1$ objects, some box must contain at least two objects. This implies $|\{q_1\alpha\} - \{q_2\alpha\}| < 1/Q$ for some $0 \leq q_1 < q_2 \leq Q$. Set

$h = m_2 - m_1$ and $k = q_2 - q_1$ where $m_1 = \lfloor q_1\alpha \rfloor$, $m_2 = \lfloor q_2\alpha \rfloor$. This gives

$$\left|\alpha - \frac{h}{k}\right| < \frac{1}{kQ}$$

with $1 \le k \le Q$. □

There are many beautiful applications of the Pigeonhole Principle. Here is one of the nicest.

Example 3.32 (Erdős and Szekeres) *Prove that every sequence of $(m-1)(n-1)+1$ distinct real numbers has either an increasing subsequence with m terms or a decreasing subsequence with n terms.*

Solution. Place an element of the sequence in a box labeled r if the longest increasing subsequence beginning with that element has r terms. If there is no increasing subsequence with m terms, we need only $m-1$ boxes. By the Pigeonhole Principle, the placement of $(m-1)(n-1)+1$ elements in these boxes yields a box with at least n elements. These n terms form a decreasing subsequence since any two terms forming an increasing subsequence belong to different boxes.

To see that this result is best possible, let $[a, b]$ denote the sequence $(a, a+1, \dots, b)$ and consider

$$I_1, I_2, \dots, I_{n-1} \quad \text{where} \quad I_k = [(n-1-k)(m-1)+1, (n-k)(m-1)].$$

This sequence consists of $(m-1)(n-1)$ distinct integers and has neither an increasing subsequence with m terms nor a decreasing subsequence with n terms. Example: for $m = 5$, $n = 4$ the sequence is

$$9, 10, 11, 12, 5, 6, 7, 8, 1, 2, 3, 4. \quad \square$$

Example 3.33 *Prove that every array of distinct real numbers with m rows and $N = (n-1)^{2^m} + 1$ columns has an $m \times n$ subarray where each row is either an increasing or a decreasing sequence.*

Solution. The case of $m = 1$ is a consequence of the result of Erdős and Szekeres. The remainder of the proof uses induction on m. Suppose we are given an array with $m > 1$ rows and $N = (n-1)^{2^m} + 1$ columns. Note that $N = (k-1)^{2^{m-1}} + 1$ where $k = (n-1)^2 + 1$. By the induction hypothesis, the array made up of the first $m-1$ rows has an $(m-1) \times k$ subarray in which every row is either increasing or decreasing. Consider the corrsponding $k = (n-1)^2 + 1$ positions in the last row. By the Erdős-Szekeres result, this sequence has either an increasing or a decreasing subsequence with n terms, and this completes the induction.

This result is best possible. We have already seen that this is the case for $m = 1$. We now argue by induction on m. Take $m > 1$ and suppose there is an array with $m-1$ rows and $p = (n-1)^{2^{m-1}}$ columns that has no $(m-1) \times n$ subarray where each row is monotone. Let the columns of this array be $u_1, u_2, \ldots, u_p$ and form a new array with $m-1$ rows and p^2 columns $cu_1 + u_1, cu_1 + u_2, \cdots, cu_1 + u_p, cu_2 + u_1, \cdots, cu_2 + u_p, \cdots, cu_p + u_p$, where c is a suitably large constant. Add one more row, namely $I_1, I_2, \ldots, I_p$ where I_k is the interval $[p(p-k) + 1, p(p-k+1)]$. The verification that this array has no $m \times n$ subarray where each row is monotone is left to the reader. □

There are many nice problems in which the Pigeonhole Principle is used to prove that under appropriate conditions a sequence of integers has a block of consecutive terms with special properties. We present two examples.

Example 3.34 *Given integers* $1 \le a_1, a_2, \ldots, a_m \le n$ *and* $1 \le b_1, b_2, \ldots, b_n$
$\le m$, *show there are integers* p, q, r, s *for which* $a_p + a_{p+1} + \cdots + a_q = b_r + b_{r+1} + \cdots + b_s$.

Solution. Suppose $a_1 + \cdots + a_m \ge b_1 + \cdots + b_n$. For $k = 1, 2 \ldots, n$, let $j = j(k)$ be the smallest index for which $a_1 + \cdots + a_j \ge b_1 + \cdots + b_k$, and set

$$c_k = (a_1 + \cdots + a_{j(k)}) - (b_1 + \cdots + b_k).$$

Since none of the a_i's exceed n and since $(a_1 + \cdots + a_{j(k)-1}) - (b_1 + \cdots + b_k) < 0$, we have $0 \le c_k < n$ for $1 \le k \le n$. If $c_k = 0$ for some k we have the desired result. Otherwise, the n numbers $c_1, c_2, \ldots, c_n$ assume only $n-1$ different values, so the Pigeonhole Principle implies $c_{r-1} = c_s$ for some $r \le s$. Then with $p = j(r-1) + 1$ and $q = j(s)$ we have

$$a_p + a_{p+1} + \cdots + a_q = b_r + b_{r+1} + \cdots + b_s.$$

The proof works the same way under the assumption $a_1 + \cdots + a_m < b_1 + \cdots + b_n$. In this case, we use the fact that none of the b_i's exceeds m. □

Example 3.35 *One of Santa's helpers makes at least one toy every day, but not more than 730 toys in a year (including leap years). Prove that for any given positive integer n the elf makes exactly n toys over some string of consecutive days.*

Note: Leap years play an essential role in this problem. Otherwise, the elf could make two toys every day and the claimed result would be false for n odd.

Solution. Consider a period consisting of k consecutive years, r of which are leap years. Then the elf works $p = 365k + r$ days to turn out at most $730k$ toys. Let a_s denote the number of toys made over the period from day 1 to day s. Consider the $2p$ numbers

$$\begin{array}{ccccc} a_1 & a_2 & a_3 & \cdots & a_p \\ a_1 + n & a_2 + n & a_3 + n & \cdots & a_p + n. \end{array}$$

Since the elf makes at least one toy every day, we have $a_1 < a_2 < \cdots < a_p$, so no two numbers in the top row are equal; likewise, no two numbers in the bottom row are equal. There are $2p = 730k + 2r$ numbers in the array, and each number is between 1 and $730k + n$. Thus, if $2r > n$ the Pigeonhole Principle guarantees that two of the numbers must be equal. We then have $a_j = a_i + n$ for some $1 \le i < j \le p$ so over days $i + 1$ through j the elf makes exactly n toys. Except for multiples of 100 that are not multiples of 400, every year divisible by 4 is a leap year. Thus, over a long enough period of time, there will be more than $n/2$ leap years, so there will be a string of consecutive days where the elf makes exactly n toys. □

Example 3.36 (1985 USAMO) *There are n people at a party. Prove that there are two people such that of the remaining $n - 2$ people, there are at least $\lfloor n/2 \rfloor - 1$ of them each of whom knows either both or neither of the two. Assume that "knowing" is a symmetric relation.*

Solution. If someone at the party knows k others, then there are $k(n - 1 - k)$ pairs where this person knows one member but not the other. There are thus $\binom{n-1}{2} - k(n - k)$ pairs where he or she knows either both members or neither one. Note that by the AM-GM inequality,

$$\binom{n-1}{2} - k(n-k) \geq \binom{n-1}{2} - \left(\frac{n-1}{2}\right)^2 = \frac{(n-1)(n-3)}{4}.$$

Introduce $\binom{n}{2}$ boxes, one for each pair of persons at the party. For each threesome $\{a, b, c\}$ at the party, place c in box $\{a, b\}$ if c knows both a and b or else neither one. Thus we have at least $n(n-1)(n-3)/4$ objects distributed among $n(n-1)/2$ boxes, so by the Pigeonhole Principle, one box has at least

$$\left\lceil \frac{n(n-1)(n-3)/4}{n(n-1)/2} \right\rceil = \left\lceil \frac{n-3}{2} \right\rceil = \left\lfloor \frac{n}{2} \right\rfloor - 1$$

objects. This gives the required conclusion. □

Note. It is appropriate to ask whether or not this result is best possible. For certain values of n (infinitely many of them, in fact) the answer is "yes." Let p be a prime congruent to 1 modulo 4, and suppose that the guests at the party are $P_0, P_1, \ldots, P_{p-1}$. Further suppose that P_i knows P_j if and only if $i-j$ is a quadratic residue (mod p). By the formula proved in §1.8, for each pair $\{i, j\}$ where $i-j$ is a quadratic residue, the number of k's for which both $k-i$ and $k-j$ are quadratic residues is $(p-5)/4$ and the number of k's for which both $k-i$ and $k-j$ are quadratic nonresidues is $(p-1)/4$. Adding these values gives $(p-3)/2 = \lfloor p/2 \rfloor - 1$. If $i-j$ is a quadratic nonresidue, these counts simply reverse roles; there are $(p-1)/4$ k's where both $k-i$ and $k-j$ are quadratic residues, and $(p-5)/4$ where both are quadratic nonresidues. Thus, for each pair at this party, the number of third persons who know either both or neither of the persons in the pair is $\lfloor p/2 \rfloor - 1$. By the appropriate generalization, this construction works for the case $n = p^\alpha \equiv 1 \pmod 4$ where p is prime and $\alpha \geq 1$.

The following three examples do not use the Pigeonhole Principle. Nevertheless, they are very close in spirit to the other examples in this section. In each case, the existence of an interesting combinatorial structure is proved by means of a quantitative argument involving inequalities.

Example 3.37 *In a circular arrangement of zeros and ones, with n terms altogether, prove that if the number of ones exceeds $(k-1)n/k$, there must be a string of k consecutive ones.*

Solution. Let $a_1, a_2, \ldots, a_n$ be the terms. Suppose r of the terms are ones and there are no k consecutive ones in the circular arrangement of these terms. Then

$$\begin{array}{ccccccccc} a_1 & + & a_2 & + & \cdots & + & a_k & \leq & k-1 \\ a_2 & + & a_3 & + & \cdots & + & a_{k+1} & \leq & k-1 \\ \vdots & \vdots & \vdots & \vdots & \cdots & \vdots & \vdots & \vdots & \vdots \\ a_n & + & a_1 & + & \cdots & + & a_{k-1} & \leq & k-1. \end{array}$$

Adding these inequalities, we get $kr \leq n(k-1)$. Hence if $r > (k-1)n/k$ there must be k consecutive ones. □

Example 3.38 *A given $m \times r$ matrix A of real numbers satisfies $a_{i1} \leq a_{i2} \leq \cdots \leq a_{in}$ for $i = 1, 2, \ldots, m$. The elements in each column are now rearranged to obtain a new matrix B satisfying $b_{1j} \leq b_{2j} \leq \cdots \leq b_{mj}$ for $j = 1, 2, \ldots n$. Prove that $b_{i1} \leq b_{i2} \leq \cdots \leq b_{in}$ for $i = 1, 2, \ldots, m$.*

Note. An alternative version of the problem may be helpful. Soldiers are in a rectangular formation of m rows and n columns, with the soldiers in each row arranged in order of increasing height from left to right. The commanding officer decides to rearrange the soldiers, one column at at time, so that each column is in order of increasing height from front to back. Show that the rows are still arranged by increasing height.

Solution. For $1 \leq j \leq n$, let

$$N_j(x) = |\{i|\ b_{ij} \geq x\}|, \quad -\infty < x < \infty.$$

Thus $N_j(x)$ counts the number of elements of the jth column of B that equal or exceed x. Since the numbers in any given column of B are the same as those in the corresponding column of A, only in a different order, and $a_{ij} \leq a_{i\,j+1}$ for $i = 1, 2, \ldots, m$, we see that for any real number x,

$$N_j(x) \leq N_{j+1}(x), \quad j = 1, 2, \ldots, n-1.$$

Suppose the desired property of B fails. Then $b_{rc} > b_{r,c+1}$ for some pair (r, c). Let $x = b_{rc}$. Then $b_{rc} \leq b_{r+1,c} \leq \cdots \leq b_{mc}$ implies $N_c(x) \geq m - r + 1$ and $b_{1,c+1} \leq b_{2,c+1} \leq b_{r,c+1} < b_{rc}$ implies $N_{c+1}(x) \leq m - r$. Since $N_c(x) > N_{c+1}(x)$ is impossible, the desired property of B holds.
□

Example 3.39 *Suppose that the squares of an $n \times n$ chessboard are labeled arbitrarily with the numbers 1 through n^2. Prove that there are two adjacent squares whose labels differ (in absolute value) by at least n.*

Solution. With each pair of adjacent squares, associate a label (min, max) where min (max) is the minimum (maximum) of the labels of the squares involved. Suppose there exist integers A, B with $A < B$ and n non-overlapping pairs of adjacent squares whose labels (a_i, b_i) satisfy

$$\max\{a_1, a_2, \ldots, a_n\} \le A, \quad B \le \min\{b_1, b_2, \ldots, b_n\}.$$

Then

$$\sum_{i=1}^{n} b_i \ge B + (B+1) + \cdots + (B+n-1) = \left(B + \frac{n-1}{2}\right)n$$

and

$$\sum_{i=1}^{n} a_i \le A + (A-1) + \cdots + (A-n+1) = \left(A - \frac{n-1}{2}\right)n,$$

so $\sum_{i=1}^{n}(b_i - a_i) \ge n^2$, from which we have $b_j - a_j \ge n$ for some j.

To see that there exist integers A, B and n pairs of adjacent squares as desired, let k be the smallest integer such that the set of labels for some row or column of the chessboard is a subset of $[k]$. (The label set for every row (column) is a subset of $[n^2]$ so such a minimum value exists.) We may assume that the square in row r and column c of the chessboard is labeled k and the labels of the other squares in this row r are all from $[k-1]$. By the minimality of k, every column except possibly column c contains a square with label $\ge k+1$. Thus in every column except possibly column c there are two adjacent squares, one with label $\le k-1$ and the other with label $\ge k+1$. If some element of column c has label $\ge k+1$, then column c contains two adjacent squares, one with label $\le k$ and the other with label $\ge k+1$. In this case, we have the desired result with $A = k$ and $B = k+1$. Otherwise, the label set for column c is a subset of $[k]$ and we have the desired situation with $A = k-1$ and $B = k$.

This result is clearly best possible since the squares of the chessboard can be labeled $1, 2, \ldots, n^2$ with the square in row i and column j receiving the label $n(i-1)+j$, which results in every pair of adjacent squares having labels that differ by either 1 or n. □

3.7 Combinatorial Averaging

It is sometimes helpful to apply the techniques of probability to combinatorial problems. We shall limit ourselves to the simplest probability model. Let S be a finite set and let $A \subseteq S$. The **probability** of A is $\text{Prob}(A) = |A|/|S|$. In particular, $\text{Prob}(\{\sigma\}) = 1/|S|$ for each $\sigma \in S$. Let $X : S \to \mathbb{R}$ be a specified function. In the present context, X is called a **random variable**. The **average** or **expected value** of X is given by

$$E(X) = \frac{1}{|S|} \sum_{\sigma \in S} X(\sigma).$$

Perhaps the most useful property of the expected value is its **linearity:** $E(cX) = cE(X)$ and

$$E(X_1 + X_2 + \cdots + X_n) = E(X_1) + E(X_2) + \cdots + E(X_n).$$

The kth **moment** of X is defined as $\mu_k = E(X^k)$, and the kth **factorial moment** is given by $E_k(X) = E(X^{\underline{k}})$. In many important applications $X = X_1 + \cdots + X_n$, where each X_i is the **characteristic function** of some $P_i \subseteq S$, namely X_i satisfies $X_i(\sigma) = 1$ if $\sigma \in P_i$ and $X_i(\sigma) = 0$ if $\sigma \notin P_i$. Then $E(X_i) = \text{Prob}(P_i)$. The kth factorial moment of X is given by

$$E_k(X) = \frac{k!}{|S|} \sum_{\sigma \in S} \binom{X(\sigma)}{k} \tag{3.20}$$

$$= \frac{k!}{|S|} \sum_{[n]^k} \sum_{\sigma \in S} X_{i_1}(\sigma) X_{i_2}(\sigma) \cdots X_{i_k}(\sigma), \tag{3.21}$$

where in (3.21) the outer sum is taken over all k-tuples of elements of $[n]$. To see that (3.20) and (3.21) yield the same value, simply note that any $\sigma \in S$ belonging to exactly r of the sets $P_1, P_2, \ldots, P_n$ contributes $\binom{r}{k}$ to each sum.

To illustrate the usefulness of these ideas in combinatorics, we begin with some easy examples.

Example 3.40 (1987 IMO) *Let $p_n(k)$ be the number of permutations of $[n]$ that have exactly k fixed points. Prove that*

$$\sum_{k=0}^{n} k \cdot p_n(k) = n!.$$

Solution. Let S_n denote the set of all permutations π of $[n]$ and let X_i be the characteristic function for the set $\{\pi |\ \pi(i) = i\}$. Then

$$\sum_{k=0}^{n} k \cdot p_n(k) = \sum_{\pi \in S_n} \sum_{i=1}^{n} X_i(\pi) = \sum_{i=1}^{n} \sum_{\pi \in S_n} X_i(\pi) = n \cdot (n-1)! = n!.$$

In probability language, $E(X) = 1$ where $X(\pi)$ is the number of fixed points of π; the expected number of fixed points in a random permutation is 1. □

Example 3.41 (Proposed for the 1985 IMO) *Find the average of the quantity*

$$(a_1 - a_2)^2 + (a_2 - a_3)^2 + \cdots + (a_{n-1} - a_n)^2$$

taken over all permutations $(a_1, a_2, \ldots, a_n)$ of $[n]$.

Solution. Consider the expected value of one term. We have

$$E((a_i - a_{i+1})^2) = \frac{(n-2)!}{n!} \sum_{r=1}^{n} \sum_{s=1}^{n} (r-s)^2 = \frac{2}{n(n-1)} \sum_{k=1}^{n-1} (n-k)k^2,$$

by realizing that for $k \neq 0$, the summand k^2 occurs $2(n-k)$ times in the double sum $\sum_{r=1}^{n} \sum_{s=1}^{n} (r-s)^2$. Thus

$$\begin{aligned} E((a_i - a_{i+1})^2) &= \frac{1}{n(n-1)} \sum_{k=1}^{n-1} [(n-k)k^2 + k(n-k)^2] \\ &= \frac{1}{n-1} \sum_{k=1}^{n-1} (n-k)k \\ &= \frac{1}{n-1} \binom{n+1}{3}, \end{aligned}$$

where the last result follows from the fact that there are $(n-k)k$ three-element subsets of $[n+1]$ where the middle term (in size) is $k+1$. (Alternatively, one can simply work out the sum from well-known formulas.) Hence, by the linearity of expectation,

$$E((a_1 - a_2)^2 + (a_2 - a_3)^2 + \cdots + (a_{n-1} - a_n)^2) = \binom{n+1}{3}. \quad \square$$

Example 3.40 was exceptionally easy (for an IMO problem). The following extension is more challenging and leads to an interesting formula.

Example 3.42 *Let $X(\pi)$ be the number of fixed points of $\pi \in S_n$. Find the moments $E(X^k)$, $(k = 1, 2, 3, \ldots)$.*

Solution. Consider the factorial moments. Note that

$$\sum_{[n]^r} \sum_{\pi \in S_n} X_{i_1}(\pi) X_{i_2}(\pi) \cdots X_{i_r}(\pi) = \binom{n}{r}(n-r)!,$$

so by (3.21) we have

$$E_r(X) = \begin{cases} 1, & r \le n \\ 0, & r > n. \end{cases}$$

In view of

$$X^k = \sum_r \left\{ {k \atop r} \right\} X^{\underline{r}},$$

we have

$$E(X^k) = \sum_{r=1}^{\min(n,k)} \left\{ {k \atop r} \right\}, \qquad k \ge 1. \quad \square$$

Note. Since there are $\binom{n}{j} D_{n-j}$ permutations of $[n]$ that have exactly j fixed points, this result can be written as

$$\sum_{j=0}^{n} \binom{n}{j} D_{n-j} j^k = n! \sum_{r=1}^{\min(n,k)} \left\{ {k \atop r} \right\}.$$

This curious combinatorial identity deserves another proof, and here is one. Let us count pairs (π, f) in which π is a permutation of $[n]$ and $f : [k] \to F(\pi)$, where $F(\pi)$ denotes the set of fixed points of π. To obtain the left-hand side, note that if $|F(\pi)| = j$ there are $\binom{n}{j} D_{n-j}$ choices for π, and for each one there are j^k choices for f. To obtain the right-hand side, note that if $|\mathcal{R}(f)| = r$ there are $\left\{ {k \atop r} \right\} \frac{n!}{(n-r)!}$ choices for f, and for each one there are $(n-r)!$ choices for π.

In preparation for the next example, let us first recall some facts about determinants. The determinant of the $n \times n$ matrix $A = [a_{ij}]$ is given by

$$\det A = \sum_{\pi} \epsilon(\pi) \prod_{i=1}^{n} a_{i,\pi(i)},$$

where the sum is taken over all permutations π of $[n]$, and $\epsilon(\pi)$ is $+1$ or -1 depending on whether π contains an even or odd number of inversions. The product of two permutations is given by composition: $(\sigma\tau)(i) = \sigma(\tau(i))$. The inverse of a permutation is defined in the obvious way: $\sigma^{-1}\sigma = \sigma\sigma^{-1} = e$, where e is the identity permutation given by $e(i) = i$, $i = 1, 2, \ldots, n$. The signature $\epsilon(\cdot)$ satisfies $\epsilon(\sigma\tau) = \epsilon(\sigma)\epsilon(\tau)$. Determinants are fairly complicated objects, so it is an especially nice fact that some results involving random determinants turn out to be simple.

Example 3.43 *Find the expected value of* $(\det A)^2$ *for a random* $n \times n$ *0–1 matrix* A.

Solution. In this case, S is the set of all $n \times n$ 0–1 matrices and $|S| = 2^{n^2}$. From the definition of the determinant,

$$\begin{aligned}(\det A)^2 &= \left(\sum_{\sigma} \epsilon(\sigma) \prod_{i=1}^{n} a_{i,\sigma(i)}\right)\left(\sum_{\tau} \epsilon(\tau) \prod_{i=1}^{n} a_{i,\tau(i)}\right) \\ &= \sum_{\sigma}\sum_{\tau} \epsilon(\sigma)\epsilon(\tau) \prod_{i=1}^{n} a_{i,\sigma(i)} a_{i,\tau(i)}.\end{aligned}$$

With σ fixed, set $\pi = \sigma^{-1}\tau$ and replace the sum over all permutations τ by the sum over π. Note that

$$\epsilon(\sigma)\epsilon(\tau) = \epsilon(\sigma^{-1})\epsilon(\tau) = \epsilon(\pi).$$

Thus we can write

$$(\det A)^2 = \sum_{\sigma}\sum_{\pi} \epsilon(\pi) \prod_{i=1}^{n} a_{i,\sigma(i)} a_{i,\sigma\pi(i)}.$$

At this point, it looks as though the calculation of $E((\det A)^2)$ might be formidable, but it turns out to be amazingly simple. Let $f(\pi) =$

$|\{i| \ \pi(i) = i\}|$ be the number of fixed points of π. Then, for each permuation σ,

$$E\left(\prod_{i=1}^{n} a_{i,\sigma(i)}a_{i,\sigma\pi(i)}\right) = \frac{2^{f(\pi)}}{2^{2n}}.$$

To see this, note that the product has $2n - f(\pi)$ distinct factors, and there are $2^{n^2-2n+f(\pi)}$ matrices in S in which all of these factors are equal to 1. Thus, by the linearity of expectation,

$$E((\det A)^2) = \frac{n!}{2^{2n}}\sum_{\pi}\epsilon(\pi)\,2^{f(\pi)}.$$

Now we observe a lovely fact: $\sum_{\pi}\epsilon(\pi)\,2^{f(\pi)}$ is itself a determinant, namely the determinant of the $n \times n$ matrix $C = [c_{ij}]$, where $c_{ij} = 2$ if $i = j$ and $c_{ij} = 1$ otherwise. This determinant is easily evaluated. Replace the first row by the sum of all the rows, and then take out a factor of $n + 1$. Now subtract the first row from each of the other rows in turn and apply the Laplace expansion. The result is $\det C = n + 1$; thus we get

$$E((\det A)^2) = \frac{(n+1)!}{2^{2n}}. \quad \square$$

Many times the process of calculating a combinatorial average can be put in terms of counting 1's in a 0–1 matrix.

Example 3.44 (1981 IMO) *Let $1 \le r \le n$ and consider all subsets of r elements of the set $[n]$. Each of these subsets has a smallest member. Let $F(n, r)$ denote the arithmetic mean of these smallest numbers; prove that*

$$F(n, r) = \frac{n+1}{r+1}.$$

Solution. With $M = \binom{n+1}{r+1}$ and $N = \binom{n}{r}$, let $\{X_1, X_2, \ldots, X_M\}$ be the collection of $(r + 1)$-element subsets of $\{0, 1, \ldots, n\}$ and let $\{Y_1, Y_2, \ldots, Y_N\}$ be the collection of r-element subsets of $[n]$. Consider the $M \times N$ 0–1 matrix in which there is a 1 in position (i, j) if the deletion of the smallest element of X_i yields Y_j and otherwise there is a 0. Each row contains a single 1, and the number of 1's in column j is the smallest element in X_j. It follows that the sum over all r-element subsets of $[n]$ of the subset's smallest element is $\binom{n+1}{r+1}$.

The arithmetic mean of the smallest elements is thus

$$F(n,r) = \binom{n+1}{r+1} \Big/ \binom{n}{r} = \frac{n+1}{r+1}. \quad \square$$

Example 3.45 *Let $X(\lambda)$ denote the number of different part sizes in a given partition λ. Show that for a random partition of $[n]$,*

$$E(X) = \frac{1}{p(n)} \sum_{k=1}^{n} p(n-k).$$

Solution. Consider the 0–1 matrix with $p(n)$ rows and n columns where the element in the row corresponding to λ and column k is 1 if $\lambda_i = k$ for some i and 0 otherwise. Then the number of 1's in the row corresponding to λ is the number of different part sizes in λ. At the same time, the number of 1's in column k is $p(n-k)$. Thus the total number of 1's is $\sum_{k=1}^{n} p(n-k)$, and the result follows. Unfortunately, there is no known simple formula for $\sum_{k=1}^{n} p(n-k)$. $\square$

Note. Let $Y(\lambda)$ denote the number of 1's in the partition λ. For $k = 1, 2, \ldots, n$, there are $p(n-k)$ partitions of n that have at least k 1's. It follows that

$$E(Y) = \frac{1}{p(n)} \sum_{k=1}^{n} p(n-k).$$

The fact that for a random partition of n the expected number of different part sizes equals the expected number of 1's was first noted by Richard Stanley.

The next example is a famous result obtained by combinatorial averaging. It is an elegant proof, given by David Lubell in 1966, of a theorem of Emanuel Sperner (1928).

Example 3.46 (Sperner's Theorem) *Suppose that $\mathcal{F}$ is a family of subsets of $[n]$ such that no subset in the family contains another one. Prove*

$$|\mathcal{F}| \le \binom{n}{\lfloor n/2 \rfloor}.$$

Solution. First note that this result is best possible; this is shown by taking $\mathcal{F}$ to be the family of all $\lfloor n/2 \rfloor$-element subsets of $[n]$. A family of subsets of $[n]$ such that no subset in the family contains another one will be called a **Sperner family**. For each permutation π of $[n]$, let $C(\pi)$ denote the chain of subsets

$$\varnothing \subset \{\pi(1)\} \subset \{\pi(1), \pi(2)\} \subset \{\pi(1), \pi(2), \pi(3)\} \subset \cdots \subset [n].$$

Given a Sperner family $\mathcal{F}$, construct a 0–1 matrix with $n!$ rows and $|\mathcal{F}|$ columns as follows. With the rows identified with the permutations of $[n]$ and the columns identified with the members of $\mathcal{F}$, the element in the row π and column A is 1 if A belongs to the chain $C(\pi)$ and 0 otherwise. Note that $A \in \mathcal{F}$ belongs to $|A|!(n-|A|)!$ chains, since if $|A| = m$ and $C(\pi)$ contains A, there are m choices for $\pi(1)$, $m-1$ choices for $\pi(2)$, and so on, while there are $n-m$ choices for $\pi(m+1)$, $n-m-1$ choices for $\pi(m+2)$, and so on. It follows that the number of 1's in the matrix is $\sum_{A\in\mathcal{F}} |A|!(n-|A|)!$. Since $\mathcal{F}$ is a Sperner family, no chain contains two or more members of $\mathcal{F}$, so there are no more than $n!$ 1's in the matrix. Hence

$$\sum_{A\in\mathcal{F}} |A|!(n-|A|) \le n!,$$

which may be written as

$$\sum_{A\in\mathcal{F}} \frac{1}{\binom{n}{|A|}} \le 1. \tag{3.22}$$

Since $\binom{n}{k} \le \binom{n}{\lfloor n/2 \rfloor}$ for $k = 0, 1, 2, \ldots, n$, we have

$$\frac{|\mathcal{F}|}{\binom{n}{\lfloor n/2 \rfloor}} \le \sum_{A\in\mathcal{F}} \frac{1}{\binom{n}{|A|}} \le 1,$$

so $|\mathcal{F}| \le \binom{n}{\lfloor n/2 \rfloor}$.

To see that this really is a case of combinatorial averaging, let S be the set of all permutations of $[n]$ and let $X(\pi) = |\mathcal{F} \cap C(\pi)|$. Then what our calculation has shown is

$$E(X) = \sum_{A\in\mathcal{F}} \frac{1}{\binom{n}{|A|}}.$$

On the other hand, the fact that $\mathcal{F}$ is a Sperner family implies $X(\pi) \le 1$ for each permutation π, so $E(X) \le 1$, and thus we have (3.22). □

Averaging arguments may be used to give nonconstructive proofs that certain combinatorial configurations exist.

Example 3.47 *Prove that if $n < \sqrt{(k-1)2^k}$ then the set of integers $[n]$ can be colored with two colors so that no k-term arithmetic progression is monochromatic.*

Solution. The k-term arithmetic progression $a,\ a+d,\ a+2d,\ \ldots,\ a+(k-1)d$ is a subset of $[n]$ if and only if $a \geq 1$ and $a + (k-1)d \leq n$. Thus, for each $a \geq 1$, there are $\lfloor (n-a)/(k-1) \rfloor$ arithmetic progressions in $[n]$ beginning with a. The total number of k-term arithmetic progressions contained in $[n]$ is therefore

$$N = \sum_{a=1}^{n-(k-1)} \left\lfloor \frac{n-a}{k-1} \right\rfloor.$$

Note that

$$N \leq \frac{1}{k-1} \sum_{r=k-1}^{n-1} r < \frac{n^2}{2(k-1)}.$$

Given a two-coloring of $[n]$ let $X = X_1 + X_2 + \cdots + X_N$ where $X_i = 1$ if the ith k-term arithmetic progression in $[n]$ is monochromatic and $X_i = 0$ otherwise. There are $2 \cdot 2^{n-k}$ colorings where $X_i = 1$, so $E(X_i) = 2^{-(k-1)}$. Thus

$$E(X) = N2^{-(k-1)} < \frac{n^2}{(k-1)2^k} < 1.$$

Since X takes on only nonnegative integer values, it follows that $X = 0$ for some coloring. This means that there is a coloring with no monochromatic k-term arithmetic progression. □

Variations of this problem have been proposed for the IMO on different occasions.

3.8 Some Extremal Problems

Many important problems in combinatorics involve finding extreme (maximum or minimum) values. We have seen several problems of this type already, and the techniques that have been used to solve them include mathematical induction, counting, and use of the Pigeonhole Principle. In this section, we point out one more technique. This involves choosing an appropriate function whose extreme values identify desired configurations.

For what is to follow, we need to use some terminology from **graph theory**. A **graph** G consists of a finite nonempty set of **vertices** $V(G)$ and a set $E(G)$ of pairs of distinct vertices, called **edges**. Thus $E(G) \subseteq \{\{u, v\} | \ u, v \in V(G)\}$. For convenience, we denote the edge $\{u, v\}$ simply as uv. If u and v are vertices and uv is an edge, we say that u and v are **neighbors** or that they are **adjacent**, and that the vertex u (or v) and the edge uv are **incident**. We shall also say that the edge uv **joins** u to v. The number of edges incident with v is called the **degree** of v and is denoted by $\deg(v)$. There is an important relation connecting the set of degrees $\{\deg(v) | \ v \in V(G)\}$ with $|E(G)|$:

$$\sum_{v \in V(G)} \deg(v) = 2|E|.$$

Since the edges are pairs of distinct vertices, a graph with n vertices has at most $\binom{n}{2}$ edges. Graphs in which every vertex pair is an edge are called **complete**. The complete graph with n vertices is denoted by K_n. A graph G is **bipartite** if there is a partition of $V(G)$ into two blocks X_1, X_2 such that no two vertices in the same block are adjacent. More generally, the graph is k-partite if there is a partition into k blocks $X_1, X_2, \ldots, X_k$ such that no two vertices in the same block are adjacent. If, in addition, any two vertices in different blocks are adjacent, the graph is called a **complete multipartite graph**. Many important questions have to do with **subgraphs**. If $V(H) \subseteq V(G)$ and $E(H) \subseteq E(G)$ we say that H is a **subgraph** of G or that G **contains** H.

Example 3.48 *Prove that the vertices of any graph can be colored using colors red and blue so that at least half of the neighbors of any red vertex are blue and at least half of the neighbors of any blue vertex are red.*

Solution. Given a coloring C, let $\Phi(C)$ denote the number of edges of the graph joining vertices of the same color. Let C_0 be a coloring for which Φ is as small as possible. We claim that for C_0 at least half of the neighbors of any red vertex are blue and at least half the neighbors of any blue vertex are red. If not, we could find a vertex with more than half of its neighbors of the same color, and change the color of this vertex to get a smaller value of Φ. □

The following problem was proposed for the 1992 IMO. It has been reworded to conform with our graph theoretic terminology.

Example 3.49 (Proposed for the 1992 IMO) *The edges of a bipartite graph G are to be colored with $n \geq 2$ colors. For such a coloring, let $\deg_i(v)$ denote the degree of vertex v in the subgraph whose edges are all those with color i. Prove that there exists a coloring such that $|\deg_i(v) - \deg_j(v)| \leq 1$ for every $v \in V(G)$ and for every pair of colors $\{i, j\}$.*

Solution. We begin with an arithmetical fact.

LEMMA. *Let*

$$\phi(d_1, d_2, \ldots, d_n) = \sum_{\{i,j\}} |d_i - d_j|,$$

where $d_1, d_2, \ldots, d_n$ are integers and the sum is over all $\binom{n}{2}$ pairs. Let $\Delta\phi$ denote the change in ϕ if the d_i's are changed in a specified way. If $d_r < d_s$ then replacing d_r and d_s by $d_r + 1$ and $d_s - 1$, respectively, yields $\Delta\phi \leq 0$. If $d_s - d_r \geq 2$, the same transformation gives $\Delta\phi \leq -2$.

Proof. By a simple calculation, $\Delta\phi = 0$ if $d_s - d_r = 1$ and $\Delta\phi = -2(k + 1)$ if $d_s - d_r > 1$, where k is the number of terms d_i such that $d_r < d_i < d_s$.

Let $V = (X, Y)$ be a partition of $V(G)$. Given a coloring of the edges of G, let

$$\Phi = \sum_{v \in V} \phi(v) \quad \text{where} \quad \phi(v) = \sum_{\{i,j\}} |\deg_i(v) - \deg_j(v)|.$$

Suppose the given coloring does not have the desired property. Then $|\deg_r(u) - \deg_s(u)| \geq 2$ for some vertex $u \in V(G)$ and some pair of colors $\{r, s\}$. Without loss of generality, we may assume that $u = x_1 \in X$ and $\deg_s(u) - \deg_r(u) \geq 2$. Choose a chain of edges

$$x_1y_1,\ y_1x_2,\ x_2y_2,\ \ldots$$

that alternates with respect to the two colors, with the first edge having color s, the second edge color r, and so on. This process stops at some point, yielding an alternating color chain starting at $u \in X$ and ending at some $z \in V$. (The chain may use vertices repeatedly, but edges occurring in the chain must be distinct.)

Suppose $z \in X$. Then $\deg_r(z) > \deg_s(z)$ since of the edges incident with z the chain has used one more edge with color r than it has with color s, and there are no more edges with color s that are incident with z since the chain cannot be extended. Exchange colors r and s along the chain. By the lemma, this transformation leads to $\Delta\phi(u) \leq -2$ and $\Delta\phi(z) \leq 0$ with $\Delta\phi(v) = 0$ for $v \neq u, z$. Hence $\Delta\Phi \leq -2$.

The analysis where $z \in Y$ is similar. In this case $d_r(z) < d_s(z)$, and exchange of colors along the chain again yields $\Delta\Phi \leq -2$.

Consider a coloring that yields the minimum possible value of Φ. In view of the procedure just described, this coloring has the property that $|\deg_i(v) - \deg_j(v)| \leq 1$ for each vertex v and each pair of colors $\{i,j\}$. □

If a graph with n vertices does not contain K_p as a subgraph, how many edges can it have? An upper bound for the special case $p = 3$ was found by W. Mantel in 1907 by an ingenious argument. Mantel's argument extends to the general case.

Example 3.50 *Prove that if G is a graph with n vertices that does not contain K_p as a subgraph, then*

$$|E(G)| \leq \frac{(p-2)n^2}{2(p-1)}.$$

Solution. Suppose that G has n vertices and contains no K_p. Assign to each vertex $v \in V(G)$ a nonnegative number $w(v)$ called its **weight** so that $\sum_{v\in V(G)} w(v) = 1$. Let

$$S = \sum_{uv\in E(G)} w(u)w(v).$$

Suppose that the w's have been chosen so as to maximize S. Then for $uv \notin E(G)$ we may assume that either $w(u) = 0$ or $w(v) = 0$. To see this, first note that without loss of generality,

$$\sum_{xu\in E(G)} w(x) \geq \sum_{zv\in E(G)} w(z).$$

If $w(u)$ and $w(v)$ were both positive, we could increase $w(u)$ by some amount, decrease $w(v)$ by the same amount, and S would not decrease. This means that the maximum value of S is attained in some case where all the positive weight vertices belong to some complete subgraph and the weights assigned to these vertices are equal (see Problem 14, §2.4). Since the largest possible complete subgraph has $p-1$ vertices, we find that

$$S \leq \binom{p-1}{2}\left(\frac{1}{p-1}\right)^2$$

for all choices of the w's. On the other hand, by assigning $w(v) = 1/n$ to each $v \in V(G)$ we get $S = |E(G)|/n^2$. It follows that

$$|E(G)| \leq \frac{(p-2)n^2}{2(p-1)}. \quad \square$$

Another ingenious proof of this result has been given by Noga Alon and Joel Spencer. This proof uses combinatorial averaging. Let $\omega(G)$ denote the order of the largest complete subgraph of G. The argument of Alon and Spencer proves that for any graph G with n vertices,

$$\omega(G) \geq \sum_{v\in V(G)} \frac{1}{n-\deg(v)}. \tag{3.23}$$

It is easy to show that this implies the upper bound on $|E|$ for a graph G with n vertices satisfying $\omega(G) \leq p-1$. If G does not contain K_p,

$$p-1 \geq \sum_{v\in V(G)} \frac{1}{n-\deg(v)} \geq \frac{n^2}{\sum_{v\in V(G)}(n-\deg(v))} = \frac{n^2}{n^2-2|E|},$$

where the second step uses the inequality

$$(a_1 + a_2 + \cdots + a_n)\left(\frac{1}{a_1} + \frac{1}{a_2} + \cdots + \frac{1}{a_n}\right) \geq n^2$$

from Chapter 2. Thus $(p-1)(n^2 - 2|E|) \leq n^2$, and

$$|E| \leq \frac{(p-2)n^2}{2(p-1)}.$$

To show that (3.23) holds, consider a random ordering π of the vertices of G. Think of the vertices as placed on a line in some random order from left to right. For each vertex $v \in V(G)$, let $X_v(\pi) = 1$ if v is adjacent to each vertex lying to its left and let $X(\pi) = \sum_{v\in V(G)} X_v(\pi)$. Including v itself, there are $n - \deg(v)$ vertices not adjacent to v, so if

$X_v = 1$ then v is the leftmost of these $n - \deg(v)$ vertices. The probability that this happens, and thus the expected value of X_v, is clearly $1/(n - \deg(v))$. Since $X(\pi) = k$ implies that for the given ordering there are k vertices each adjacent to every vertex to its left, we have $X(\pi) \leq \omega(G)$. Thus $E(X) \leq \omega(G)$, so by the linearity of expectation,

$$\omega(G) \geq E(X) = \sum_{v \in V(G)} E(X_v) = \sum_{v \in V(G)} \frac{1}{n - \deg(v)}. \quad \square$$

Let us prove that the bound $|E| \leq (p-2)n^2/(2(p-1))$ is best possible. Suppose that n is a multiple of $p-1$ and consider the complete multipartite graph with $p-1$ parts and $n/(p-1)$ vertices in each part. By the Pigeonhole Principle, any set of p vertices contains two vertices in the same part, which must then be nonadjacent. Thus this graph contains no K_p. Moreover, it has n vertices and

$$\binom{p-1}{2}\left(\frac{n}{p-1}\right)^2 = \frac{(p-2)n^2}{2(p-1)}$$

edges.

The following problem from the 1995 USAMO provides yet another proof that in a graph with n vertices that contains no triangle (K_3) the number of edges is at most $n^2/4$.

Example 3.51 *Suppose that in a certain society each pair of persons can be classified as either* **amicable** *or* **hostile**. *We shall say that each member of an amicable pair is a* **friend** *of the other and each member of a hostile pair is a* **foe** *of the other. Suppose that the society has n persons and q amicable pairs, and that at least one pair out of every set of three persons is hostile. Prove that there is at least one member of the society whose foes include $q(1 - 4q/n^2)$ or fewer amicable pairs.*

Solution. In graph theoretic terms, the problem is to prove that in a triangle-free graph G with n vertices and q edges there is at least one vertex v_0 such that the number of edges joining pairs of vertices distinct from and not adjacent to v_0 is at most $q(1-4q/n^2)$. For future reference, note that the relation

$$\sum_{w \in V} \deg(w) = 2q$$

together with Cauchy's inequality implies

$$n\sum_{w\in V}(\deg(w))^2 \geq \left(\sum_{w\in V}\deg(w)\right)^2 = 4q^2. \qquad (3.24)$$

For $v \in V(G)$ let $f(v) = \sum_{w\in N(v)} \deg(w)$. Since G is triangle-free, the number of edges joining pairs of vertices distinct from and not adjacent to v is $q - f(v)$. Thus the problem reduces to proving that there is a vertex v_0 such that $f(v_0) \geq (2q/n)^2$. Using (3.24), we find that the average value of f satisfies

$$\begin{aligned}
\bar{f} &= \frac{1}{n}\sum_{v\in V} f(v) \\
&= \frac{1}{n}\sum_{v\in V}\sum_{w\in N(v)} \deg(w) \\
&= \frac{1}{n}\sum_{w\in V} \deg(w) \sum_{v\in N(w)} 1 \\
&= \frac{1}{n}\sum_{w\in V}(\deg(w))^2 \\
&\geq 4q^2/n^2.
\end{aligned}$$

Thus such a vertex exists. □

Since the quantity of concern in this problem cannot be negative, we have $q \leq n^2/4$ for any triangle-free graph with n vertices and q edges.

We have seen that $|E(G)| \leq (p-2)n^2/(2(p-1))$ for any graph with n vertices that does not contain K_p as a subgraph, and that this result is best possible when n is a multiple of $p-1$. This leaves open the case in which n is not a multiple of $p-1$. What is the best possible result in this case? The solution to this problem was found by Paul Turán in 1941. First let us see what the answer should be. Based on our previous experience, we suspect that the critical example should be a multipartite graph with $p-1$ parts. To maximize the number of edges, we want the parts to be as nearly equal in size as possible. Write $n = k(p-1) + r$ where $0 \leq r < p-1$, and consider the multipartite graph with $p-1-r$ parts of size k and the remaining

r parts of size $k+1$. This graph has

$$M(n,p) = \frac{(p-2)n^2}{2(p-1)} - \frac{r(p-1-r)}{2(p-1)}$$

edges. This form is instructive since it shows the "correction" to the upper bound in Example 3.50.

Example 3.52 (Turán) *Prove that if G has n vertices and contains no K_p then*

$$|E(G)| \le \frac{(p-2)n^2}{2(p-1)} - \frac{r(p-1-r)}{2(p-1)}.$$

Furthermore, equality holds if and only if G is the Turán graph, namely the complete multipartite graph with $p-1-r$ parts of size $k = (n-r)/(p-1)$ and r parts of size $k+1$.

Solution. The proof is by induction on $k = \lfloor n/(p-1) \rfloor$. The case $k = 0$ corresponds to $n = r \le p-1$, and in this case it is obvious that the maximum number of edges is $\binom{n}{2}$, with equality holding when the graph is complete. Now assume $k \ge 1$ and let G be a graph with $n = k(p-1)+r$ vertices containing no K_p and having the maximum possible number of edges subject to this condition.

Note that G must contain K_{p-1} as a subgraph; otherwise we could add an edge, and the resulting graph would still not contain K_p. Pick such a complete subgraph with $p-1$ vertices, let X be its vertex set, and let $Y = V(G) - X$. Let G_Y denote the subgraph of G **induced** by Y; this is the graph whose vertex set is Y and whose edges are all the edges of G joining pairs of vertices in Y. Now G_Y has $n-p+1 = (k-1)(p-1)+r$ vertices, and by induction, the number of edges in G_Y satisfies

$$\begin{aligned}
|E(G_Y)| &\le M(n-p+1,p) \\
&= \frac{(p-2)}{2(p-1)}(n-p+1)^2 - \frac{r(p-1-r)}{2(p-1)} \\
&= M(n,p) - (p-2)(n-p+1) - \binom{p-1}{2}
\end{aligned}$$

edges. Since G contains no K_p, no vertex in Y is adjacent to all the vertices of X. It follows that the total number of edges in G satisfies

$$\begin{aligned}|E(G)| &\le |E(G_Y)| + (p-2)(n-p-1) + \binom{p-1}{2}\\ &\le M(n,p),\end{aligned}$$

completing the proof by induction.

To obtain equality, (i) G_Y must have $M(n-p+1,p)$ edges and (ii) each vertex in Y must be joined to $p-2$ vertices of X. An inductive argument then shows that the only graph with n vertices and $M(n,p)$ edges not having K_p as a subgraph is the Turán graph. □

We have now come full circle. The last proof demonstrates again the power of mathematical induction, one of the first techniques discussed in this book. This is a good place to end.

Exercises for Sections 3.6-3.8

1. Prove that for any set of $n+1$ integers chosen from $[2n]$, one element of the set divides another one.
2. Prove that for any nine lattice points in three-dimensional space there is a segment joining two of the points that contains another lattice point.
3. Show that for any collection of n integers (not necessarily distinct) there is a nonempty subcollection whose sum is divisible by n. *Note.* A collection in which elements may be repeated is called a **multiset**. Various important combinatorial problems can be naturally expressed in terms of this concept.
4. Let $A = \{a_1, a_2, \ldots, a_n\}$ be a set of distinct positive integers, and let $c_1, c_2, \ldots, c_N$ be a sequence of length N whose terms are elements of A. Prove that if $N \ge 2^n$ there must be a block of consecutive terms of the sequence whose product is a perfect square. Also prove that this is not necessarily the case when $N \le 2^n - 1$.
5. Given a multiset of $n+1$ positive integers with sum $2n$, prove that the multiset can be split into two parts, each with sum n.
6. Suppose that 22 points are chosen from the 7×7 grid $G = \{(i,j) | \ 1 \le i,j \le 7\}$. Prove that four of the chosen points are the vertices of a rectangle (with horizontal and vertical sides).

Also, show that this is not necessarily the case if 21 points are chosen.

7. Prove that the expected number of blocks in a random partition of $[n]$ is $(B_{n+1} - B_n)/B_n$, where

$$B_m = \sum_k \left\{ {m \atop k} \right\}$$

is the total number of partitions of $[m]$. This is the mth **Bell number**.

8. For any nonempty set S of numbers let $\sigma(S)$ and $\pi(S)$ denote the sum and product, respectively, of the elements of S. Prove that

$$\sum \frac{\sigma(S)}{\pi(S)} = (n^2 + 2n) - \left(1 + \frac{1}{2} + \frac{1}{3} + \cdots + \frac{1}{n}\right)(n + 1),$$

where $\sum$ denotes the sum over all nonempty subsets S of $[n]$. (1991 USAMO)

9. Given a 0–1 matrix with $m \geq 2$ rows, no two identical, having at least m columns, prove there is a column whose deletion yields a matrix with no two identical rows.

10. An $n \times n$ chessboard has some of its squares colored red and the other squares colored black. If a black square has at least two of its four sides in common with red squares, you are permitted to change the color from black to red. Suppose that by repeating this operation as long as it is allowed, all of the squares of the chessboard are eventually colored red. Prove that there were at least n red squares in the original chessboard.

11. You are given two partitions of $[n]$, the first with k blocks and the second with $k + m$ blocks. Prove that there are at least $m + 1$ elements of $[n]$ that are in smaller blocks in the second partition than in the first.

12. Show that the edges of any graph with n vertices can be covered using $\lfloor n^2/4 \rfloor$ or fewer edges (K_2's) and triangles (K_3's). *Note.* In a covering, each edge of the given graph occurs in *at least one* of the covering graphs. In this case, a stronger result is true. The edges of any graph with n vertices can be partitioned using

$\lfloor n^2/4 \rfloor$ or fewer edges and triangles. Thus each edge occurs in *exactly one* of the partitioning graphs. The reader is encouraged to try to prove the stronger result.

13. Denote by $\|z_i - z_j\|$ the usual Euclidean distance between two points $z_i = (x_i, y_i)$ and $z_j = (x_j, y_j)$ in the plane:
$$\|z_i - z_j\| = \sqrt{(x_i - x_j)^2 + (y_i - y_j)^2}.$$
Let $z_1, z_2, \ldots, z_{3p}$ be points in the plane such that $\|z_i - z_j\| \leq 1$ for all i, j. Prove that at most $3p^2$ of the distances are greater than $\sqrt{2}/2$.
14. Prove that for $n \geq 5$, every graph with n vertices and $\lfloor n^2/4 \rfloor + 2$ edges contains two triangles with exactly one vertex in common (a **bowtie**). [Proposed (with different wording) for the 1988 IMO]
15. A **chain** is a sequence of edges in a graph so that consecutive edges have a vertex in common. Given a graph with n vertices and q edges numbered $1, 2, \ldots, q$, show that there exists a chain of at least $\lceil 2q/n \rceil$ edges that is monotonic with respect to the edge numbering. [Proposed for the 1986 IMO]

Olympiad Problems for Chapter 3

1. Prove that
$$\sum_{k=0}^{995} \frac{(-1)^k}{1991 - k}\binom{1991 - k}{k} = \frac{1}{1991}.$$
[Proposed for the 1991 IMO]
2. A positive integer is called **evil** if the number of digits 1 in its binary expansion is even. For example $18 = (10010)_2$ is evil. Find the sum of the first 1985 evil numbers. [1985 British Olympiad, Further International Selection Test]
3. Seventeen people correspond by mail with one another - each one with all the rest. In their letters only three different topics are discussed. Each pair of correspondents deals with only one of these topics. Prove that there are at least three people who write to each other about the same topic. [1964 IMO]
4. For any permutation π of $[n]$ let $d(\pi)$ denote the sum
$$|\pi(1) - 1| + |\pi(2) - 2| + \cdots + |\pi(n) - n|,$$

and let $i(\pi)$ denote the number of inversions of π. In other words, $i(\pi)$ counts the pairs $1 \leq i < j \leq n$ such that $\pi(i) > \pi(j)$. Prove that $i(\pi) \leq d(\pi)$. [1991 Czechoslovak Mathematical Olympiad]

5. Nine mathematicians at an international conference discover that among any three of them, at least two speak a common language. If each of the mathematicians can speak at most three languages, prove that there are at least three of the mathematicians who can speak the same language. [1978 USAMO]

6. A certain organization has n members, and it has $n + 1$ three-member committees, no two of which have identical membership. Prove that there are two committees that share exactly one member. [1979 USAMO]

7. A difficult mathematical competition consisted of a Part I and a Part II with a combined total of 28 problems. Each contestant solved 7 problems altogether. For each pair of problems, there were exactly two contestants who solved both of them. Prove that there was a contestant who, in Part I, solved either no problems or else at least four problems. [1984 USAMO]

8. A function $f(S)$ assigns to each nine-element subset S of the set $\{1, 2, \ldots, 20\}$ a whole number from 1 to 20. Prove that regardless of how the function f is chosen, there will be a ten-element subset $T \subset \{1, 2, \ldots, 20\}$ such that $f(T - \{k\}) \neq k$ for all $k \in T$. [1988 USAMO]

9. Let $S = \{1, 2, 3, \ldots, 280\}$. Find the smallest integer n such that each n-element subset of S contains five numbers that are pairwise relatively prime. [1991 IMO]

10. On an infinite chessboard, a game is played as follows. At the start, n^2 pieces are arranged on the chessboard in an $n \times n$ block of adjoining squares, one piece in each square. A move in the game is a jump in a horizontal or vertical direction over an adjacent occupied square to an unoccupied square immediately beyond. The piece that has been jumped over is then removed. Find the values of n for which the game can end with only one piece remaining on the board. [1993 IMO]

Hints and Answers for Selected Exercises

Section 1.1

1. No, it is not possible. The product of any two members of the progression $5, 9, 13, 17, 21, \cdots$ also belongs to this sequence, while M belongs to the progression $3, 7, 11, 15, 19, \cdots$
2. Note that $n^4 + 4 = (n^2 + 2n + 2)(n^2 - 2n + 2)$. $n = 1$.
3. Note that $2^8 + 2^{11} = 2^8 3^2 = 48^2$. Thus if $2^8 + 2^{11} + 2^n = k^2$, then $2^n = (k - 48)(k + 48)$ and there are nonnegative integers a and b such that $k - 48 = 2^a$ and $k + 48 = 2^b$. $n = 12$.
4. $7744 = 88^2$.
5. Note that $3(2n + 3) - 2(3n + 4) = 1$.
6. Suppose that $(n-2)^2 + (n-1)^2 + n^2 + (n - 1)^2 + (n + 2)^2 = m^2$. Then $5n^2 + 10 = m^2$, and it follows that m is divisible by 5 and thus $n^2 + 2$ is divisible by 5. Write $n = 5q - r$ where $0 \leq r \leq 4$. Since $n^2 + 2$ is divisible by 5, so is $r^2 + 2$. But $r^2 + 2 = 2, 3, 6, 11, 18$ for $r = 0, 1, 2, 3, 4$, respectively.
7. If $(n-1)n(n+1) = n(n^2-1)$ is a perfect kth power where $k \geq 2$, then n^2 and $n^2 - 1$ must be kth powers, and this is impossible.
8. $N = 1156$.

9. Note that $2(a^2+b^2) = (a+b)^2 + (a-b)^2$ and $(a^2+b^2)(c^2+d^2) = (ac-bd)^2 + (ad+bc)^2$.

10. Say that n is a **score** if it can be expressed as $xa + yb$ where x and y are nonnegative integers. Since $(a, b) = 1$, there are integers x, y such that $xa + yb = 1$, and it follows that for every $n \in \mathbb{Z}$ there are unique corresponding integers x, y such that $0 \le x < b$ and $n = xa + yb$. It follows that the largest number that is not a score is $M = (b-1)a + (-1)b$. Consider the set of $(a-1)(b-1)$ numbers $S = \{0, 1, 2, \ldots, M\}$. Show that $m \in S$ is a score if and only if $M - m$ is not a score. Thus, exactly $(a-1)(b-1)/2$ natural numbers are not scores. $a = 8$, $b = 5$.

11. Show that $\lfloor x + y \rfloor \ge \lfloor x \rfloor + \lfloor y \rfloor$ and use this fact together with Legendre's formula to show that $\alpha_p(m+n) \ge \alpha_p(m) + \alpha_p(n)$. Thus $(m+1)(m+2)\cdots(m+n)$ is divisible by $n!$.

12. Write $n = 2^k + r$ where $0 \le r < 2^k$. Then $\alpha_2(n!) \le n/2 + n/4 + \cdots + n/2^k = (1 - 2^{-k})n \le n - 1$ with equality if and only if $r = 0$. $n = 2^k$, $k = 0, 1, 2, \ldots$

13. In the sequence $1, 4, 7, \ldots, 1000$, there are (i) 67 terms divisible by 5, (ii) 14 terms divisible by 5^2, (iii) 3 terms divisible by 5^3, and (iv) 1 term divisible by 5^4. The exponent of 5 in the prime factorization of $1 \cdot 4 \cdot 7 \cdots 1000$ is thus $67 + 14 + 3 + 1 = 85$. This is also the number of terminal zeros in the decimal representation.

14. $20! = 2^{18} \cdot 3^8 \cdot 5^4 \cdot 7^2 \cdot 11 \cdot 13 \cdot 17 \cdot 19$.

15. Let $r = \lfloor \log n / \log 2 \rfloor$. Then $2^r \le n < 2^{r+1}$ and $e_2(k) \le r - 1$ for all $1 \le k \le n$ except $k = 2^r$. Multiply both sides of the equation

$$H_n = 1 + \frac{1}{2} + \frac{1}{3} + \cdots + \frac{1}{n}$$

by $2^{r-1}\mathrm{LCM}(1, 3, \ldots, 2\lfloor (n-1)/2 \rfloor + 1)$. Then every term on the right-hand side is an integer *except one*.

Section 1.2

3. $x_n = 2^{n-1} + 1$.

4. $1^2 + 3^2 + \cdots + (2n-1)^2 = (2n-1)(2n)(2n+1)/6$.

5. $1^4 + 2^4 + \cdots + n^4 = n(n+1)(2n+1)(3n^2+3n-1)/30$.

6. Inductive step: $(1 + x)^{n+1} = (1 + x)(1 + x)^n \geq (1 + x)(1 + nx) = 1 + (n + 1)x + nx^2 \geq 1 + (n + 1)x$.

7. Inductive step: $|(x_1 + x_2 + \cdots + x_n) + x_{n+1}| \geq |x_1 + \cdots + x_n| + |x_{n+1}| \geq |x_1| + |x_2| + \cdots + |x_n| + |x_{n+1}|$.

8. For $n = 2$ the inequality is equivalent to $(1 - x_1)(1 - x_2) \geq 0$. This holds for $0 \leq x_1, x_2 \leq 1$ with equality if and only if either $x_1 = 1$ or $x_2 = 1$. To make the inductive step, use the result for $n = 2$ as well as the one for $n - 1$. Thus

$$\begin{aligned}(1 + x_1)\cdots(1 + x_{n+1}) &\leq 2^{n-1}(1 + x_1x_2\cdots x_n)(1 + x_{n+1})\\ &\leq 2^n(1 + x_1x_2\cdots x_{n+1}).\end{aligned}$$

9. By the Euclidean Algorithm, $(F_{n+1}, F_n) = (F_n, F_{n-1})$.

10. Note that $a + 1 = a^2$ and $b + 1 = b^2$. Inductive step:

$$\begin{aligned}F_{n+1} &= F_n + F_{n-1}\\ &= \frac{1}{\sqrt{5}}[(a^n - b^n) + (a^{n-1} - b_{n-1})]\\ &= \frac{1}{\sqrt{5}}[(a + 1)a^{n-1} - (b + 1)b^{n-1}]\\ &= \frac{1}{\sqrt{5}}[a^{n+1} - b^{n+1}].\end{aligned}$$

11. Inductive step: $F_1^2 + F_2^2 + \cdots + F_n^2 + F_{n+1}^2 = F_nF_{n+1} + F_{n+1}^2 = F_{n+1}(F_n + F_{n+1}) = F_{n+1}F_{n+2}$.

12. Let $r = (7 + \sqrt{37})/2$ and $s = (7 - \sqrt{37})/2$. These are the roots of $x^2 - 7x + 3 = 0$. Then

$$a_{n+2} = r^{n+2} + s^{n+2} = r^n(7r - 3) + s^n(7s - 3) = 7a_{n+1} - 3a_n.$$

To prove that $a_n - 1$ is divisible by 3 for every n, write $a_{n+2} - 1 = 7(a_{n+1} - 1) - 3(a_n - 2)$ and use induction. Since $0 < s < 1$ and $a_n = r^n - s^n$ is an integer, it follows that $a_n = r^n + s^n = \lfloor r^n \rfloor + 1$.

13. The $n = 1$ case is trivial. Label the vertices $P_1, P_2, \ldots, P_{2n+1}$ in clockwise order. Claim: there are two nonintersecting diagonals of the form P_iP_{i+2} whose endpoints have different colors. To see this, first consider the colors of $P_1, P_3, \ldots, P_{2n+1}$. Since P_1 and P_{2n+1} have different colors, we have at least one such diagonal. If there are two, we are done, so assume there is just one, say

$P_{2k-1}P_{2k+1}$. Then P_1, P_{2k}, and P_{2n+1} have different colors. Consider the colors of $P_1, P_{2n}, P_{2n-2}, \ldots, P_2$. If we find two diagonals as desired, there is nothing more to prove; if not, P_2 has the same color as P_{2k}, and the diagonals $P_{2k-1}P_{2k+1}$ and $P_{2n+1}P_2$ fulfill our requirement. Delete the appropriate triangles of the form $P_iP_{i+1}P_{i+2}$ and apply induction.

14. The case of a 3×3 checkerboard may be dealt with by exhaustion, using symmetry to limit the number of cases. If one of the dimensions of the checkerboard (C) is at least five, there are two non-overlapping strips of width two on opposite sides of the board. If one of these strips (S) contains none of the three deleted squares, then the induction is immediate. Otherwise we may assume that S contains exactly one of the deleted squares and $C - S$ contains the other two. If we choose any associated square $C - S$ that has the same color as the deleted square in S, then we can argue by induction that with the associated square deleted, $C - S$ can be covered with dominoes. By making an appropriate choice of the associated square, extend the covering to C.

15. By symmetry, we may assume that x_n is the largest term among $x_1, \ldots, x_n$. Then $(x_{n-1} + x_1)/x_n$ is either 1 or 2. Note that $(x_{n-1} + x_1)/x_n = 2$ can be satisfied only in case $x_1 = x_2 = \cdots = x_n$. In case $(x_{n-1} + x_1)/x_n = 1$, delete x_n from the circle and apply induction. Equality holds in case $x_k = k$, $k = 1, 2, \ldots, n$.

Section 1.3

2. Three applications of Fermat's little thorem yield

$$20^{15} \equiv 9^{15} \equiv 27^{10} \equiv 1 \pmod{11},$$
$$20^{15} \equiv 144^{15} \equiv 12^{30} \equiv 1 \pmod{31},$$
$$20^{15} \equiv 81^{15} \equiv 3^{60} \equiv 1 \pmod{61}.$$

3. (a) Since $\phi(44) = 20$ and $19 \equiv -25 \pmod{44}$, Euler's theorem yields $19^{10} \equiv 5^{20} \equiv 1 \pmod{44}$. Also $69 \equiv -19 \pmod{44}$.
   (b) Note that $2^6 \equiv -1 \pmod{13}$ and that $3^3 \equiv 1 \pmod{13}$.
4. 929.
5. (a) $n^5 - n = (n-1)n(n+1)(n^2+1)$ is divisible by 6.

(b) $(2k+1)^5 - (2k+1) = 8(2k+1)k(k+1)(2k^2+2k+1)$ is divisible by 16.

8. The square of any integer is congruent to 0, 1, 2, or 4 (mod 7).
9. 2^k, where $k \equiv 5 \pmod 8$.
10.

$$(n^3+1, n^2+2) = \begin{cases} 9, & n \equiv 5 \pmod 9, \\ 3, & n \equiv 2 \text{ or } 8 \pmod 9, \\ 1, & \text{otherwise.} \end{cases}$$

11. (a) $n = 3k$, $k = 0, 1, 2, \ldots$

    (b) Show that $2^n + 1$ is congruent to 2, 3, 5 (mod 7) for n congruent to 0, 1, 2 (mod 3), respectively.
12. 721.
13. 2519; $2520k - 1$, $k = 1, 2, 3, \ldots$
14. $x \equiv 77 \pmod{120}$.

Section 1.4

1. First consider the case in which 143 immediately follows the decimal point. Show that the assumption $a/b = .143\ldots$, where a and b are integers and $0 < b < 100$, leads to the absurd conclusion that $7a - b$ is an integer between 0 and 1. Show that the general case can be reduced to the one in which 143 immediately follows the decimal.
2. (a) Show that $x = \sqrt{2} + \sqrt[3]{3}$ satisfies $x^6 + c_1x^5 + \cdots + c_5x + 1 = 0$, where $c_1, \ldots, c_5$ are integers.

    (b) If $\log_{10} 2 = a/b$ then $10^a = 2^b$.
3. Using

$$\cos 3\theta = 4\cos^3\theta - 3\cos\theta$$
$$\cos 4\theta = 8\cos^4\theta - 8\cos^2\theta + 1$$

together with $\cos(4\pi/7) + \cos(3\pi/7) = 0$ leads to the quartic equation $x^4 + x^3 - 4x^2 - 3x + 2 = 0$, satisfied by $x = \cos(\pi/7)$. Note that $x = -2$ is a root. Reduction leads to the cubic equation $x^3 - x^2 - 2x + 1 = 0$. Another approach is to start from the fact that $z = \cos(\pi/7) + i\sin(\pi/7)$ satisfies $z^7 + 1 = 0$. The fact that

$\cos(\pi/7)$ is irrational is now an easy consequence of the Rational Root Theorem.

4. If $1 + \sqrt{3} = (a + b\sqrt{3})^2$, where a and b are rational, then $a^2 + 3b^2 = 1$ and $2ab = 1$. But then $a^2 + 3b^2 = 2ab$, which yields $(a-b)^2 + 4b^2 = 0$, and this is satisfied by real a, b if and only if $a = b = 0$. $(5 + 2\sqrt{3})^2 = 37 + 20\sqrt{3}$.

5. Note that if $d_s d_{s+1} = d_r d_{r+1}$ for some $s > r$, then the decimal expansion of x is periodic and hence x is rational. Such a repetition is inevitable since there are only 100 possibilities for a pair of consecutive digits.

6. $\sqrt{a} = \dfrac{8as^2 + (s^2 - a - b - c)^2 - 4(ab + bc + ca)}{4s(s^2 + a - b - c)}$, where $s = \sqrt{a} + \sqrt{b} + \sqrt{c}$.

7. If $np \neq 0$ then $\sqrt{6} = (m^2 - 2n^2 - 3p^2)/(2np)$. If $n \neq 0$, $p = 0$ then $\sqrt{2} = -m/n$, and if $n = 0$, $p \neq 0$ then $\sqrt{3} = -m/p$.

8. Comparing coefficients in

$$(x - u)(x - v)(x - uv) = x^3 + ax^2 + bx + c,$$

we have $-a = u + v + uv$, $b = uv(1 + u + v)$, $c = -u^2v^2$. Thus $(1 - a)uv = b - c$, and if $a \neq 1$ then $uv = (b - c)/(1 - a)$.

9. $\left|\dfrac{r + 2}{r + 1} - \sqrt{2}\right| = \dfrac{(\sqrt{2} - 1)|r - \sqrt{2}|}{r + 1}$.

Section 1.5

1. $|z + w|^2 + |z - w|^2 = (z + w)(\overline{z} + \overline{w}) + (z - w)(\overline{z} - \overline{w}) = 2\{|z|^2 + |w|^2\}$. The sum of squares of the lengths of the diagonals of a parallelogram equals the sum of squares of the four sides.

2. $\{i, (-\sqrt{3} - i)/2, (\sqrt{3} - i)/2\}$.

3. (a) $z = 3/4 + i$.

   (b) $z = 0, 1 - i$.

4. $z = (-3 \pm 6\sqrt{2})(1 + i)$.

5. If $3c_2 = c_1^2$ then the equation can be written in the form

$$\left(z + \frac{c_1}{3}\right)^2 = \left(\frac{c_1}{3}\right)^3 - c_3.$$

Thus if a is a cube root of $c_1^3/27 - c_3$, the roots of the equation are given by

$$z + \frac{c_1}{3} = a,\ a\omega,\ a\omega^2,$$

and it follows that the corresponding points in the complex plane are vertices of an equilateral triangle. The above steps can be reversed to prove the converse. Note that $3c_2 = c_1^3$ is equivalent to

$$z_1^2 + z_2^2 + z_3^2 = z_1z_2 + z_2z_3 + z_3z_1.$$

6. Note that $abc(\overline{a} + \overline{b} + \overline{c}) = r^2(bc + ac + ab)$. Thus

$$\left|\frac{ab + bc + ca}{a + b + c}\right| = \frac{|abc|}{r^2} = r.$$

7. $t = i(1 - z)/(1 + z)$.
8. For $r = 0, 1, 2,$

$$\sum_{k \equiv r \ (\mathrm{mod}\ 3)} \binom{n}{k} = \frac{1}{3}\left[2^n + 2\cos\left(\frac{(n - 2r)\pi}{3}\right)\right].$$

9. Evaluate the real part of both sides:

$$\sum_{k=0}^{n} \binom{n}{k} (e^{i\theta})^k = (1 + e^{i\theta})^n.$$

10. $\{\pm\sqrt{2},\ (\sqrt{3} \pm i)/2,\ -(\sqrt{3} \pm i)/2\}$.
11. Note that z is a root of the given equation if and only if

$$z = \frac{1 + 2w}{1 - 2w},$$

where w is a fifth root of unity. Thus

$$\left|z + \frac{5}{3}\right| = \frac{4}{3}\left|\frac{2 - w}{1 - 2w}\right| = \frac{4}{3}\left|\frac{2\bar{w} - 1}{1 - 2w}\right| = \frac{4}{3}.$$

12. $z = i\tan\left(\dfrac{(2k - 1)\pi}{4n}\right), \qquad k = 1, \ldots, 2n.$

Section 1.6

1. $(10^{n+1} - 9n - 10)/81$.

2. 36200000001.
3. Since $b^2 = (a^2 + c^2)/2$, it follows that $(b+c)(a+b) = (a+c)(a + 2b + c)/2$.
4. $20n^2$.
5. To prove the first identity, note that the sum may be written in telescoping form

$$\frac{1}{d}\left[\left(\frac{1}{a_1} - \frac{1}{a_2}\right) + \cdots + \left(\frac{1}{a_{n-1}} - \frac{1}{a_n}\right)\right].$$

The second identity is similar; the sum may be written in telescoping form

$$\frac{1}{d}\left[(\sqrt{a_2} - \sqrt{a_1}) + \cdots + (\sqrt{a_n} - \sqrt{a_{n-1}})\right].$$

6. If $a + (p-1)d = q$ and $a + (q-1)d = p$ where $p \neq q$, then $a = p + q - 1$ and $d = -1$. It follows that $a_{p+q} = 0$.
7. $n\left[a_1 a_n + \dfrac{2n-1}{6(n-1)}(a_n - a_1)^2\right]$.
8. Set

$$n\left(\frac{2a_1 + (n-1)2}{2}\right) = n^k.$$

Then $a_1 = n(n^{k-2} - 1) + 1$ is an odd integer. The progression $n^{k-1} - n + 1, n^{k-1} - n + 3, \ldots, n^{k-1} + n - 1$ fulfills the requirement.

10.

$$\begin{aligned}\sum_{k=1}^{n} \frac{F_{k+1}}{F_k F_{k+2}} &= \sum_{k=1}^{n}\left(\frac{1}{F_k} - \frac{1}{F_{k+2}}\right)\\ &= 1 + 1 - \frac{1}{F_{n+1}} - \frac{1}{F_{n+2}}\\ &= 2 - \frac{F_{n+3}}{F_{n+1}F_{n+2}}.\end{aligned}$$

11.

$$\begin{aligned}\sum_{k=1}^{n} \frac{1}{F_{2^k}} &= \frac{1}{F_1} + \frac{1}{F_2} + \sum_{k=2}^{n}\left(\frac{F_{2^{k-1}-1}}{F_{2^{k-1}}} - \frac{F_{2^k-1}}{F_{2^k}}\right)\\ &= 3 - \frac{F_{2^n-1}}{F_{2^n}}.\end{aligned}$$

12. Note that

$$\frac{k}{k^4 + k^2 + 1} = \frac{1}{2}\left[\frac{1}{(k-1)k+1} - \frac{1}{k(k+1)+1}\right],$$

so the sum telescopes. $n(n+1)/(2(n^2 - n + 1))$.

13. Using the relation

$$\binom{n}{k} = \binom{n-1}{k-1} + \binom{n-1}{k},$$

write the sum in telescoping form. $(-1)^r\binom{n-1}{r}$.

14. $a_n = 3 \cdot 2^{n-1} - n - 1$.

15. The sum is the real part of

$$\sum_{k=0}^{n-1}(k+1)(e^{2\pi i/n})^k.$$

Use

$$\sum_{k=0}^{n-1}(k+1)z^k = \frac{1 - nz^n}{1 - z} + \frac{z - z^r}{(1-z)^2}, \qquad z \neq 1.$$

Section 1.7

1. 84 years.
2. $x = z(z+1)/2,\ y = z(z-1)/2$.
3. Write the equation as $(x+1)(x^2+1) = 2^k$ and so deduce that $x + 1 = 2^r$ and $x^2 + 1 = 2^s$, where r and s are nonnegative integers. Show that $x^2 + 1 = 2^s$ has a solution only for $s = 0$ and $s = 1$. $(x, y) = (0, 0), (1, 2)$.
4. Write the equation as a quadratic in x: the discriminant must be a perfect square. $(x, y) = (0, 0), (1, -1), (-1, 1)$.
5. Multiply the equation by 5 and complete the square to obtain $(5m - 3n)^2 + 26n^2 = 9940$. Thus if (m, n) is a solution then $(5m - 3n)^2 \equiv 8 \pmod 3$.
6. Consider the equation (mod 8).
7. Since $b \neq 0$, we have $2b^2 + (a^2 - 3)b + 3a^2 + a = 0$. Using the discriminant condition, show that $a(a - 8)$ must be a perfect square. $(a, b) = (-1, -1), (8, -10), (9, -6), (9, -21)$.

8. Consider the equation (mod 16). There are no solutions.
9. For $x \geq 4$, the equation

$$y^2 = 1 + 2 + 6 + 24 + \cdots + x!$$

requires that $y^2 \equiv 3 \pmod 5$, which is impossible. $(x, y) = (1, 1), (3, 3)$.
10. Consider the equation (mod 3). $(a, b, c) = (0, 0, 0)$.
11. If $2 + 2\sqrt{28n^2 + 1}$ is an integer, then $28n^2 + 1 = (2m + 1)^2$ for some nonnegative integer m. This leads to $7n^2 = m(m + 1)$. Since $(m, m + 1) = 1$, the only possibilities are (i) $m = a^2$, $m + 1 = 7b^2$ and (ii) $m = 7c^2$, $m + 1 = d^2$. However $7b^2 - a^2 = 1$, which requires that $a^2 \equiv -1 \pmod 7$, is impossible. Thus $2 + 2\sqrt{28n^2 + 1} = 4(m - 1) = 4d^2$.
12. If x is even, then $x^4 + 4^x$ is even and $x^4 + 4^x \neq p$. If $x = 2k - 1$, then

$$\begin{aligned} x^4 + 4^x &= (x^2 + 2^x)^2 - (x\,2^k)^2 \\ &= (x^2 + x\,2^k + 2^x)(x^2 - x\,2^k + 2^x). \end{aligned}$$

13. Note that the equation reduces to $y(y + 1) = z(z + 2)$, where $z = x(x + 1)$. $(x, y) = (0, 0), (0, -1), (-1, 0), (-1, -1)$.
14. $x = (2r - 1)/(1 - r^2)$, where r is a rational number in the interval $(-1, 1)$.
15. Show that the equation can be written in the form

$$(2n + 1)^2 - 8m^2 = 1.$$

Use the standard method to solve this Pell equation. $(2n + 1) + m\sqrt{8} = (3 + \sqrt{8})^k$, $k = 1, 2, 3, \ldots$

Section 2.1

1. $2^{20} - 1$.
2. $c = 2$.
3. $n \equiv 1 \pmod 6$.
4. $R(x) = x$.
5. $k = 5/6$.
6. $a = F_{16} = 987$.

7. $\{(m, n) \mid n \equiv 1 \pmod{m+1}\}$.
8. $(a-b)(b-c)(c-a)(a+b+c)$.
9. $7ab(a+b)(a^2+ab+b^2)^2$.
10. Letting $x = a + b$, $y = b + c$, $z = c + a$, the expression to be factored becomes

$$(x+y+z)^3 - x^3 - y^3 - z^3,$$

and this expression vanishes if $x + y = 0$, $y + z = 0$, or $z + x = 0$. We have

$$(x+y+z)^3 - x^3 - y^3 - z^3 = 3(x+y)(y+z)(z+x).$$

Thus the given expression factors as follows:

$$3(2a+b+c)(a+2b+c)(a+b+2c).$$

11. $P(x) = \dfrac{1 + (-1)^n x(x-1)\cdots(x-n)/(n+1)!}{x+1}$.
12.

$$\begin{aligned} P(n+1) &= \operatorname{Im}\left\{(e^{i\pi/n})^{n+1} - (e^{i\pi/n} - 1)^{n+1}\right\} \\ &= \sin\frac{\pi}{n} + (-1)^k\left(2\sin\frac{\pi}{2n}\right)^{n+1}\begin{cases} \sin\frac{\pi}{2n}, & n = 2k, \\ \cos\frac{\pi}{2n}, & n = 2k+1. \end{cases} \end{aligned}$$

13. Use the fact that if $P(k) = r^k$ for $k = 0, 1, \ldots, n$, where P is a polynomial of degree at most n, then

$$P(n+1) = r^{n+1} - (r-1)^{n+1}.$$

By a simple extension of this result, if $P(k) = r^k$ for $k = m, m+1, \ldots, m+n$, then

$$P(m+n+1) = r^m\left\{r^{n+1} - (r-1)^{n+1}\right\}.$$

Suppose that $P(k) = F_k = (a^k - b^k)/\sqrt{5}$ for $k = m, m+1, \ldots, m+n$, where $m \geq n+2$. Use the preceding result together with $a - 1 = 1/a$ and $b - 1 = 1/b$ to conclude that

$$P(m+n+1) = F_{m+n+1} - F_{m-n-1}.$$

In particular, for $m = 992$ and $n = 990$, we have

$$P(1983) = F_{1983} - F_1 = F_{1983} - 1.$$

14. Show that the first and third equations imply that $4\sqrt[6]{y} + 4\sqrt[6]{z} + \sqrt[6]{yz} = 0$. Thus $(x, y, z) = (2^{12}, 0, 0)$ is the only solution.

15. $\displaystyle\sum_{j=1}^{n} \frac{x_j}{2j+1} = 1 - \frac{1}{(2n+1)^2}$.

Section 2.2

1. (a) $\{(1 \pm \sqrt{5})/2,\ (1 \pm i\sqrt{11})/2\}$.
   (b) $\{(1 \pm \sqrt{21})/2,\ (1 \pm i\sqrt{3})/2\}$.
   (c) $\{1,\ 3,\ 2 \pm \sqrt{6}\}$.
2. (a) $\{(3 \pm \sqrt{13})/2,\ (3 \pm i\sqrt{3})/2\}$.
   (b) $\{1/2,\ (-1 \pm i3\sqrt{3})/4\}$.
3. $2 \pm \sqrt{3}$.
4. $1 + \sqrt{2},\ \pm i\sqrt{2}$.
5. $a \geq 27/4$.
6. $\{-1,\ (3 \pm \sqrt{5})/2,\ (1 \pm i\sqrt{3})/2\}$.
7. $\{-1 \pm i\sqrt{3},\ 1 \pm 3i\}$.
8. $x^4 + y^4 + z^4 = 29$.
9. $x^4 + y^4 = 81,\ -567/2$.
10. Let a, b, c be the roots of the equation

$$x^3 + px^2 + qx + r = 0.$$

Newton's formulas yield $S_1 = -p = 0$, $S_2 = -2q$, $S_3 = -3r$, $S_5 = 5qr$. Thus $6S_5 = 5S_2S_3$, which is equivalent to the stated identity.

11. Let

$$P(x) = x^3 + ax^2 + bx + c = (x - x_1)(x - x_2)(x - x_3)$$

and let $Q(x)$ denote the desired polynomial. Then

$$Q(x^3) = (x^3 - x_1^3)(x^3 - x_2^3)(x^3 - x_3^3) = P(x)P(\omega x)P(\omega^2 x).$$

Using the identity

$$(r + s + t)(r + \omega s + \omega^2 t)(r + \omega^2 s + \omega t) = r^3 + s^3 + t^3 - 3rst$$

and setting $r = x^3 + c$, $s = ax^2$, $t = bx$, we obtain

$$Q(x^3) = (x^3 + c)^3 + a^3x^6 + b^3x^3 - 3(x^3 + c)abx^3.$$

Thus

$$Q(x) = x^3 + (a^3 - 3ab + 3c)x^2 + (b^3 - 3abc + 3c^2)x + c^3.$$

12. $A^2 = s(s^3 + as^2 + bs + c)$, where $s = -a/2$. Thus

$$A = \frac{1}{4}\sqrt{a(4ab - a^3 - 8c)}.$$

13. Express the condition

$$\left(x_1 - \frac{x_2 + x_3}{2}\right)\left(x_2 - \frac{x_1 + x_3}{2}\right)\left(x_3 - \frac{x_1 + x_2}{2}\right) = 0$$

in terms of a, b, c.

14. $\{-5, -2, 1, 4, 7\}$.

15. $S_1 = -1$; $S_2 = S_3 = \cdots = S_n = 0$.

Section 2.3

1. If $(x - a)(x - b)(x - c) \neq 0$, the equation is equivalent to

$$(a + b + c)x^2 - (ab + bc + ca)x = 0.$$

Thus $x = 0$ is a solution since $abc \neq 0$. Since $a + b + c \neq 0$,

$$x = \frac{ab + bc + ca}{a + b + c}$$

is also a solution if

$$(bc - a^2)(ca - b^2)(ab - c^2) \neq 0.$$

If $(bc-a^2)(ca-b^2)(ab-c^2) = 0$, then $x = 0$ is the only solution.

2. $x = (1 + \sqrt{5})/2$.

3. $\{0, 1\}$.

4. $x = 1 - 3\left\{\sqrt[3]{(1 + \sqrt{5})/2} - \sqrt[3]{(1 - \sqrt{5})/2}\right\}$.

5. In view of the identity

$$\sqrt{x + 1} - \sqrt{x} = \frac{1}{\sqrt{x + 1} + \sqrt{x}},$$

it is clear that the equation has a real solution only if $0 < a \leq 1$. The solution is then given by

$$x = \frac{1}{4}\left(\frac{1}{a} - a\right)^2.$$

6. $x = 1/81$.
7. $\left\{8,\ 8 + \frac{12\sqrt{21}}{7},\ 8 - \frac{12\sqrt{21}}{7}\right\}$.
8. $\{2,\ 3\}$.
9. $\{(1 + \sqrt{3})/2,\ \sqrt{2}/2\}$.
10. $\{2,\ 3,\ (-5 + \sqrt{41})/2,\ (-5 - \sqrt{33})/2\}$.
11. Let $R(x) = |x+1| - x + 4$ be the right side of the equation. The left side is $L_+(x) = 2x - a,\ x \geq a/2$ and $L_-(x) = -2x + a,\ x \leq a/2$. The equation $L_+(x) = R(x)$ has the unique solution $x = (a+5)/2$ if $a \geq -7$ and the unique solution $x = (a + 3)/4$ if $a < -7$. The equation $L_-(x) = R(x)$ has the unique solution $x = (a - 5)/2$ if $a > 3$, the solution set $\{x|\ x \leq -1\}$ if $a = 3$, and no solution if $a < 3$.
12. $4 - \sqrt{12} \leq c \leq (-7 + \sqrt{45})/2$.
13. $0 \leq x < 1$.
14. The inequality is satisfied if $x \leq (-5 - \sqrt{165})/14$ or $x \geq (-5 + \sqrt{165})/14$.
15. If $a \leq 1$ then the inequality is satisfied by every real number x. If $a > 1$, the solution set is

$$\left\{x \,\middle|\, x < \frac{2a-1}{2(a-1)}\right\}.$$

Section 2.4

1. The expression on the left is $\left\{\left(a - \frac{1}{2}\right)^2 + \frac{3}{4}\right\}\{a^4 + 1\}$.

2. $\left(\frac{a}{b} - \frac{b}{c}\right)^2 + \left(\frac{b}{c} - \frac{c}{a}\right)^2 + \left(\frac{c}{a} - \frac{a}{b}\right)^2 \geq 0.$

3. Set $x_1 = a + 1/a$, $x_2 = b + 1/b$, and $y_1 = y_2 = 1$. Then Cauchy's inequality gives

$$\left(a + \frac{1}{a}\right)^2 + \left(b + \frac{1}{b}\right)^2 \geq \frac{1}{2}\left(1 + \frac{1}{ab}\right)^2,$$

and we are practically finished.

4. The inequality can be written

$$a^2b^2c^2(ab + bc + ca) \leq a^8 + b^8 + c^8.$$

Use $ab + bc + ca \leq a^2 + b^2 + c^2$ and the power mean inequality.

5. Since $x(1 - x) \leq 1/4$ for $0 \leq x \leq 1$,

$$a(1-b) \cdot b(1-c) \cdot c(1-a) = a(1-a) \cdot b(1-b) \cdot c(1-c) \leq 1/64,$$

so at least one of the three factors is $\leq 1/4$

6. Set $x_k = (S - a_k)/S$ for $k = 1, 2, \ldots, n$ and use the inequality

$$\left(\frac{1}{x_1} + \cdots + \frac{1}{x_n}\right)(x_1 + \cdots + x_n) \geq n^2.$$

7. Let $x_1 = (a + b + c)/(b + c)$, $x_2 = (a + b + c)/(a + c)$, $x_3 = (a + b + c)/(a + b)$ and apply the above inequality.

8. If A is the area of the triangle, then

$$A^2/s = (s - a)(s - b)(s - c) \leq (s/3)^3$$

by AM-GM.

9. $4\pi\sqrt{3}R^3/9$.

10. If a, b, c, x, y, z are real numbers that satisfy $ax + by + cz = 2$, then

$$(a^2 + b^2 + c^2)(x^2 + y^2 + z^2) \geq 4$$

by Cauchy's inequality. Let $s = a^2 + b^2 + c^2$. Then $s > 0$, and by a simple application of AM-GM, we find that

$$a^2 + b^2 + c^2 + 3(x^2 + y^2 + z^2) \geq s + \frac{12}{s} \geq 2\sqrt{12}.$$

Since $2\sqrt{12} > 6$, the given system has no real solution.

11. By the AM-GM inequality, we get $1 + a^4 \geq 2a^2$, and so on. Also $(a^4b^2 + b^4c^2 + c^4a^2) \geq 3a^2b^2c^2$.

15. It follows from the AM-GM inequality that $\sigma_k \geq \binom{n}{k}$, where σ_k denotes the kth elementary symmetric function of $a_1, \ldots, a_n$.

Sections 3.2-3.5

1. (a) Count (two different ways) the number of r-member committees that can be selected from the members of an organization with m men and n women.

   (b) Let A and B be two disjoint sets with $|A| = m$ and $|B| = n$. Count (two different ways) the number of ordered pairs (X, Y) where $X \subseteq A$, $Y \subseteq B \cup X$, and $|Y| = m$.

2. (a) Deleting the first term (A) and the last term (B) from such a sequence defines a bijection onto the set of all ballot sequences with $n-1$ A's and $n-1$ B's. Thus there are C_{n-1} sequences where a tie is attained only at the end.

   (b) $C_0C_{n-2} + C_1C_{n-3} + \cdots + C_{n-2}C_0 = C_{n-1}$.

4. Let a_n denote the total number of allowed strings of length n, and let b_n, c_n, d_n denote the number of allowed strings starting with 0, 1, 2, respectively. There are then b_n strings starting with 4 and c_n strings starting with 3. Show that $a_n = 2b_n + 2c_n + d_n$, $b_n = c_{n-1}$, $c_n = b_{n-1} + d_{n-1}$, $d_n = 2c_{n-1}$, and thus show that $c_n = 3c_{n-2}$, with $c_1 = 1$ and $c_2 = 2$. Thus, $a_1 = 1$, and

$$a_{2n} = 8 \cdot 3^{n-1}, \quad a_{2n+1} = 14 \cdot 3^{n-1}, \quad n \geq 1.$$

5. Let a_n denote the value of the sum. The generating function for (a_n) is

$$G(x) = F(x) + F^2(x) + F^3(x) + \cdots = \frac{F(x)}{1 - F(x)},$$

where $F(x) = x/(1-x-x^2)$ is the Fibonacci generating function. Thus $G(x) = x/(1 - 2x - x^2)$, and by partial fractions, $a_n = \sqrt{2}((1 + \sqrt{2})^n - (1 - \sqrt{2})^n)/4$.

6. Counting the empty string as satisfying the condition, the generating function is

$$G(x) = \frac{1}{(1-x)(1-x-x^2)}.$$

There are $F_{n+3} - 1$ such strings.

7. The mapping $b_i = a_i + i - 1$ associates with each k-element alternating set $\{a_1, a_2, \ldots, a_k\} \subseteq \{k, k+1, \ldots, n\}$ a corresponding set $\{b_1, b_2, \ldots, b_k\} \subseteq \{k, k+1, \ldots, n+k-1\}$ in which each element is odd. The number of alternating fat subsets of $[n]$ is

$$\sum_{k\geq 0}^{\frac{n}{2}} \binom{\frac{n}{2}}{k} = 2^{n/2}$$

if n is even, and it is

$$\sum_{k\geq 0} \binom{\frac{n-1}{2}}{2k} + \sum_{k\geq 0} \binom{\frac{n+1}{2}}{2k+1} = \begin{cases} 2, & n = 1 \\ 3 \cdot 2^{(n-3)/2}, & n > 1 \end{cases}$$

if n is odd. This problem also has a nice solution by generating functions.

8. Map $(x_1, x_2, \ldots, x_n)$ to $(0, y_1, \ldots, y_n)$ by setting $y_i = y_{i-1}$ if $x_i = 0$ and $y_i \neq y_{i-1}$ if $x_i = 1$. This mapping is a bijection from the set of all 0-1 strings of length n with no k consecutive zeros onto the set of all 0-1 strings $(0, y_1, y_2, \ldots, y_n)$ with no $k+1$ consecutive zeros and no $k+1$ consecutive ones. Thus there are a_n of these strings. There are another a_n strings $(1, y_1, y_2, \ldots, y_n)$, and the result follows.

9. This formula can be proved by induction. Another way is to note that the sum in question equals $[x^n]F^2(x)$, where $F(x)$ is the generating function for Fibonacci numbers. Calculus can be used to advantage here. Apply term-by-term differentiation to

$$\sum_{n=0}^{\infty} F_{n+1}x^n = \frac{1}{1-x-x^2} \quad \text{and} \quad \sum_{n=1}^{\infty} F_n x^{n+1} = \frac{x^2}{1-x-x^2}.$$

10. The generating function for (a_n) is $G(x) = (1-x)^{-1/2}$, and thus $a_n = 4^{-n}\binom{2n}{n}$.

11. Use generating functions:

$$\prod_{k=1}^{\infty} (1 + x^k + \cdots + x^{k(d-1)}) = \prod_{k=1}^{\infty} \frac{1-x^{kd}}{1-x^k}.$$

12. Use generating functions:

$$\prod_{k=1}^{\infty} \frac{1 + x^{2k}}{1 - x^{2k-1}} = \prod_{k=1}^{\infty} \frac{1 - x^{4k}}{1 - x^k} = \prod_{k=1}^{\infty}(1 + x^k + x^{2k} + x^{3k}).$$

13. There is a simple combinatorial proof. Given a self-conjugate partition with Durfee square of size d, view its Ferrers diagram as consisting of d L-shaped pieces, the kth one consisting of $\mu_k = 2r_k + 1$ dots, where r_k is the number of parts greater than k. The corresponding partition $(\mu_1, \mu_2, \ldots, \mu_d)$ has d distinct parts, each one being odd.

14. $n! \sum_{k=0}^{\lfloor n/r \rfloor} \frac{(-1)^k}{k! r^k}$.

15. Use the inclusion-exclusion formula to count the number of 0-1 strings of length n that have exactly r zeros followed immediately by ones.

Sections 3.2-3.5

1. Place an integer from the given set in box B_k if its largest odd divisor is $2k - 1$ ($k = 1, 2, \ldots, n$). By the Pigeonhole Principle, at least one box has two or more occupants.

2. There must be two points, say (a_1, a_2, a_3) and (b_1, b_2, b_3), such that $a_i \equiv b_i \pmod 2$, $i = 1, 2, 3$.

3. Let $\{a_1, a_2, \ldots, a_n\}$ be the multiset of integers, and consider the sums $a_1, a_1 + a_2, \ldots, a_1 + a_2 + \cdots + a_n$. If one of these is divisible by n we have the desired result. If not, only $n - 1$ residue classes are represented, so by the Pigeonhole Principle, two of the sums are congruent modulo n.

4. Construct a 0-1 matrix with N rows and n columns in which the element in row i and column j is 0 or 1 if the number of times a_j occurs in the sequence $c_1, c_2, \ldots, c_i$ is even or odd, respectively. If a row of this matrix consists entirely of zeros, we have the desired result. If not, there are at least 2^n rows, but only $2^n - 1$ distinct possibilities for these rows.

5. It is convenient to prove a stronger result, namely, if M is any multiset of positive integers with $|M| = n + 1$ and $\sigma(M) = 2n$, then for each positive integer $k \leq 2n$ there exists a submultiset

K of M such that $\sigma(K) = k$. Let M be a multiset as specified and let $\nu(a)$ denote the multiplicity of a in M. Then we claim that if $m > 1$ is the largest element of M then $\nu(1) \geq m$. To see this, note that since $n + 1 - \nu(1)$ elements of M exceed 1 and one element equals m, we have $2n = \sigma(M) \geq \nu(1) + (n - \nu(1))2 + m$. Delete m 1's from M to obtain M', and consider the set of numbers obtained by evaluating $\sigma(K)$ for each submultiset K of M'. The resulting numbers are between 1 and $2n - m$. In general there will be gaps; in other words, the values of $\sigma(K)$ as K ranges over M' do not yield all of the numbers $1, 2, \ldots, 2n - m$. However, since no element of M' is greater than m, no gap will be greater than m. By using the m 1's, we can fill in all of the gaps.

6. For $i = 1, 2, \ldots, 7$, let a_i denote the number of points chosen from row i. Then

$$\sum_{i=1}^{7} \binom{a_i}{2} \geq 7\binom{22/7}{2} > \binom{7}{2}$$

by Jensen's inequality. (The function

$$f(x) = \begin{cases} 0, & x < 1, \\ x(x-1)/2, & x \geq 1 \end{cases}$$

is convex.) The following set shows that 21 points do not suffice:

$$\begin{array}{ccccccc} (1,1) & (1,2) & (1,3) & (2,3) & (2,4) & (2,5) & (3,1) \\ (3,5) & (3,6) & (4,1) & (4,4) & (4,7) & (5,2) & (5,5) \\ (5,7) & (6,3) & (6,6) & (6,7) & (7,2) & (7,4) & (7,6). \end{array}$$

7. Using the recurrence formula for Stirling numbers of the second kind, we obtain

$$\sum_k k \left\{ {n \atop k} \right\} = \sum_k \left(\left\{ {n+1 \atop k} \right\} - \left\{ {n \atop k-1} \right\} \right) = B_{n+1} - B_n.$$

8. Set $\sigma(\varnothing) = 0$ and $\pi(\varnothing) = 1$. Then $\sum \sigma(S)/\pi(S)$ can be extended over all subsets of $[n]$ without changing the result. Note that

$$\sum_{S \subseteq [k]} \frac{1}{\pi(S)} = \prod_{j=1}^{n} \left(1 + \frac{1}{j}\right) = k + 1.$$

Let $A_k = \sum_{S\subseteq[k]} \sigma(S)/\pi(S)$. Then using the above result, we obtain

$$A_k - A_{k-1} = \sum_{S\subseteq[k-1]} \frac{\sigma(S)+k}{k\pi(S)} = \frac{1}{k}A_{k-1} + k.$$

Write the recurrence formula as

$$\frac{A_k}{k+1} - \frac{A_{k-1}}{k} = \frac{k}{k+1}.$$

This holds for all $k \geq 1$ with $A_0 = 0$. Summing from $k = 1$ to $k = n$, we obtain

$$\frac{A_n}{n+1} = \sum_{k=1}^{n}\left(1 - \frac{1}{k+1}\right),$$

and thus $A_n = (n^2 + 2n) - (n+1)H_n$, where $H_n = 1 + 1/2 + \cdots + 1/n$.

9. We use induction on m. The result holds for $m = 2$ since for a 0-1 matrix with two rows and at least two columns, there is a column in which the two rows differ, and the deletion of any other column gives the desired result. Now suppose $m > 2$ and let A be the given matrix. Delete the first column to obtain A'. If A' has no two identical rows, there is nothing more to prove. Otherwise, the identical rows in A' come in pairs that are distinguished in A by their entries in the first column; one has a 0 and the other has a 1 in this position. Let A'' be the matrix whose rows are the *distinct* rows of A'. Then A'' has $k < m$ rows, at least k columns, and no two of its rows are identical. By induction, there is a column that can be deleted from A'' to yield a matrix with no two identical rows. The deletion of this column from A yields the desired result.

10. Given a colored chessboard C, let $M(C)$ denote the total perimeter of the red portion of the board; thus $M(C)$ is the number of edges belonging to exactly one red square. If all of the squares in the $n \times n$ chessboard are red, then M counts the edges on the boundary of the board, so its value is $4n$. Note that each time the operation is performed, M either decreases or remains the same. Thus the initial board C_0 satisfies $M(C_0) \geq 4n$, and this implies that initially there are at least n red squares.

11. For $i = 1, 2, \ldots, n$, suppose i is in a block of size a_i in the first partition and in a block of size b_i in the second partition. Then
$$\sum_{i=1}^{n} \frac{1}{a_i} = k \quad \text{and} \quad \sum_{i=1}^{n} \frac{1}{b_i} = k + m, \quad \text{so} \quad \sum_{i=1}^{n} \left(\frac{1}{b_i} - \frac{1}{a_i}\right) = m.$$
Each term in the last sum belongs to the open interval $(-1, 1)$, so at least $m + 1$ of its terms are positive.

12. This is obviously true for $n = 2$ and $n = 3$. Use induction, where the inductive step goes from n to $n + 2$. Note that $\lfloor (n + 2)^2/4 \rfloor - \lfloor n^2/4 \rfloor = n + 1$. Given a graph G with $n + 2$ vertices, let uv be an edge. (If there is no edge, there is nothing to prove.) By induction, the graph obtained by deleting u and v can be covered with $\lfloor n^2/4 \rfloor$ or fewer edges and triangles. For each of the n vertices $w \neq u, v$, if there are edges joining w to u and/or v, these can be covered with either one edge or one triangle. One edge may be necessary to cover uv. Thus at most $\lfloor n^2/4 \rfloor + n + 1 = \lfloor (n + 2)^2 \rfloor$ edges and triangles are needed to cover the edges of G.

13. Construct a graph in which the vertices represent the $3p$ points, and vertex i is adjacent to vertex j if $\|z_i - z_j\| > \sqrt{2}/2$. This graph does not contain a K_4.

14. Say that a graph with n vertices and q edges is a (n, q) graph. Let $f(n) = \lfloor n^2/4 \rfloor + 2$. It is easy to check that every $(5, 8)$ graph has a bowtie. We use induction to prove that for $n \geq 5$ every $(n, f(n))$ graph has a bowtie. Given a $(n, f(n))$ graph with $n > 5$, the deletion of a vertex of smallest degree yields a $(n-1, f(n-1))$ graph, *except* for the case $n = 7$, and then only if the graph in question is one in which every vertex has degree 4.

15. For each vertex v in the graph, let $L(v)$ denote the length of the longest chain beginning with v that is decreasing with respect to the edge numbering. To prove that $L(v) \geq 2n/q$ for some vertex v, it suffices to show that $\sum_v L(v) \geq 2q$. This is done by induction on q (with the vertex set of the graph being fixed). The claim is obviously true for $q = 1$ since if v_1v_2 is the single edge, $L(v_1) = L(v_2) = 1$. Assume that the claim is true for graphs with $q - 1$ edges and let G be a graph with the specified vertex set and q edges numbered $1, 2, 3, \ldots, q$. Remove the edge, say

v_1v_2, numbered q. Let G' denote the resulting graph, and let L' play the role for G' that L plays for G. Then clearly $L(v_1) \geq L'(v_1) + 1$, $L(v_2) \geq L'(v_2) + 1$, and $L(v) = L'(v)$ for $v \neq v_1, v_2$. Thus, by induction,

$$\sum_v L(v) = 2 + \sum_v L'(v) \geq 2 + 2(q-1) = 2q.$$

General References

1. G. Andrews, *Number Theory*, W. B. Saunders, Philadelphia, 1971.
2. G. Andrews, *The Theory of Partitions*, Addison-Wesley, Reading, Massachusetts, 1976.
3. E. J. Barbeau, *Polynomials*, Springer-Verlag, New York, 1989.
4. B. Bollobás, *Graph Theory, An Introductory Course*, Springer-Verlag, New York, 1979.
5. D. Cohen, *Basic Techniques of Combinatorial Theory*, John Wiley & Sons, New York, 1978.
6. L. Comptet, *Advanced Combinatorics*, Riedel, Dordrecht, 1974.
7. R. L. Graham, N. Patashnik, and D. Knuth, *Concrete Mathematics*, Addison-Wesley, Reading, Massachusetts, 1989.
8. S. L. Greitzer, *International Mathematical Olympiads 1959-1977*, Mathematical Association of America, Washington, D.C., 1978.
9. H. S. Hall and S. R. Knight, *Higher Algebra*, Macmillan, London, 1932.
10. G. H. Hardy, J. E. Littlewood, and G. Polya, *Inequalities*, 2nd ed., Cambridge University Press, Cambridge, 1952.

11. G. H. Hardy and E. M. Wright, *An Introduction to the Theory of Numbers*, 5th ed., Oxford University Press, Oxford, 1979.

12. M. S. Klamkin, *International Mathematical Olympiads 1979-1985*, Mathematical Association of America, Washington, D.C., 1986.

13. M. S. Klamkin, *USA Mathematical Olympiads 1972-1986*, Mathematical Association of America, Washington, D.C., 1988.

14. L. Larson, *Problem Solving Through Problems*, Springer-Verlag, New York, 1983.

15. C. L. Liu, *Introduction to Combinatorial Mathematics*, McGraw-Hall, New York 1968.

16. L. Lovász, *Combinatorial Problems and Exercises*, North-Holland, Amsterdam, 1979.

17. Z. A. Melzak, *Companion to Concrete Mathematics*, John Wiley & Sons, New York, 1973.

18. D. S. Mitrinović, *Analytic Inequalities*, Springer-Verlag, Heidelberg, 1970.

19. I. Niven, H. S. Zuckerman, and H. L. Montgomery, *An Introduction to the Theory of Numbers*, 5th ed., John Wiley & Sons, New York, 1991.

20. G. Pólya, *How to Solve It*, Doubleday, New York, 1957.

21. H. E. Rose, *A Course in Number Theory*, Oxford University Press, Oxford, 1994.

22. W. Sierpinski, *250 Problems in Elementary Number Theory*, American Elsivier, New York, 1970.

23. I. Tomescu, *Problems in Combinatorics and Graph Theory*, John Wiley & Sons, New York, 1985.

24. A. Tucker, *Applied Combinatorics*, John Wiley & Sons, New York, 1980.

25. J. Uspensky, *Theory of Equations*, McGraw-Hill, New York, 1948.

26. H. S. Wilf, *Generatingfunctionology*, Academic Press, San Diego, 1990.

List of Symbols

$\mathbb{Z}$	set of integers
$\mathbb{Z}_p$	integers modulo p
$\mathbb{R}$	set of real numbers
$\mathbb{R}^n$	n-dimensional Euclidean space
$\mathbb{C}$	set of complex numbers
$\varnothing$	empty set
$a \in S$	a is an element of the set S
$A \subseteq B$	A is a subset of B
$A \cup B$	union of sets
$A \cap B$	intersection of sets
$\|S\|$	number of elements of the set S
$A \times B$	Cartesian product of sets
$\Rightarrow$	implies

$a \mid b$	a divides b
(a, b)	greatest common divisor of a and b
$\lfloor x \rfloor$	floor, greatest integer $\leq x$
$\lceil x \rceil$	ceiling, least integer $\geq x$
$\{x\}$	fractional part of x; $\{x\} = x - \lfloor x \rfloor$
$a \equiv b \pmod{m}$	a congruent to b modulo m
$\phi(n)$	Euler's totient function
$\overline{z}$	complex conjugate of z
$\lvert z \rvert$	modulus of z; $\lvert z \rvert^2 = z\overline{z}$
$\operatorname{Re} z$	real part of z
$\operatorname{Im} z$	imaginary part of z
$x^{\underline{n}}$	falling factorial, $x(x-1)\cdots(x-n+1)$
$x^{\overline{n}}$	rising factorial, $x(x+1)\cdots(x+n-1)$
$\binom{n}{k}$	binomial coefficient
$\left(\frac{a}{p}\right)$	Legendre symbol
$\left[{n \atop k}\right]$	Stirling number of the first kind
$\left\{{n \atop k}\right\}$	Stirling number of the second kind

Index

A

Abel, Niels Henrik, 92, 164
Algebraic equations, 106–11
Al-Khowârizmî, 73
Alon, Noga, 206
André, Antoine Désiré, 146
Archimedes, 64
Argand, Jean Robert, 37
Arithmetic progression, 47

B

Ballot problem, 145
Basics of counting, 142–148
Beatty's theorem, 34
Bell numbers, 211
Bernoulli, Jakob
  inequality, 17, 125
  polynomial, 51
Bernoulli, Johann, 35
Binet's formula, 151
Binomial theorem, 40–41

C

Catalan numbers, 146
Chebyshev, Pafnuty L.
  inequality, 130–131
  polynomials, 95
  theorem, 8
Chinese remainder theorem, 23–24
Classical inequalities, 113–136
Combinatorial
  averaging, 195–202
  proof, 143–144
Complex numbers, 35–44
Compositions, 164
Conditional inequalities, 111–112
Congruence, 18–27

Convex functions, 119–120
Convex sets, 154
Cramer's rule, 90

D

De Moivre, Abraham, 35
  theorem, 36–37
Derangements, 152
Descartes, René, 35
Determinant, 89
  Laplace expansion for, 89
  of a random matrix, 198–199
Diophantine equations, 56–64
Diophantus, 56
Dirichlet, P. G. Lejeune, 188
Division algorithm, 74
Durfee square, 169
Dyson, Freeman, 92

E

Euclid, 2
  algorithm, 4
  lemma, 2–3
Euler, Leonhard, 35
  phi function, 10
  theorem, 21
Extremal problems, 202–210

F

Factor theorem, 74–75
Factorial
  falling, 181
  rising, 160
Fermat, Pierre de
  last theorem, 56–57, 61
  little theorem, 20
Ferrari, Ludovico, 92
  method, 96–97
Ferrers diagram, 167
Ferro, Scipione del, 92
Feynman, Richard, 92
Fibonacci sequence, 17
Fundamental theorem
  algebra, 39–40
  arithmetic, 3
  symmetric functions, 100–101

G

Galois, Évariste, 92
Gauss, Carl Friedrich
  complex plane, 37
  congruence, 18–19
  elimination method, 84–85
  fundamental theorem of algebra, 39–40
  quadratic reciprocity, 65
  17-gon construction, 39
  as a student, 46
Generating functions, 156–178
Geometric progression, 49
Gleason, Andrew, 129
Goldbach, Christian, 3
Graph, 203
  bipartite, 203
  complete, 203
Greatest common divisor, 2

H

Hadamard, Jacques, 8
Hardy, Godfrey Harold, 165
Harmonic number, 11
Hilbert matrix, 90

I

Inclusion-exclusion, 178–186
Inequalities
  AM-GM, 114
  Bernoulli, 125
  Cauchy, 115
  Chebyshev, 130–131
  Hölder, 124, 126
  Jensen, 120
  Minkowski, 124–125
  power mean, 114
  rearrangement, 128
  triangle, 115
  weighted AM-GM, 121
Infinite descent, 61
Interpolation, 80–84
  Lagrange's formula, 83
  Newton's method, 83–84
Irrational numbers, 29–33
Ivory, James, 21

K

Kedlaya, Kiran, 127–128

L

Legendre, Adrien Marie,
  formula, 6
  symbol, 65
Leibnitz, Gottfried Wilhelm, 141
Lubell, David, 200
Lurie, Jacob, 176

M

Mantel, W., 205
Mathematical induction, 11–18
Matrix, 84
Multiplicative function, 9
Multisection formula, 43

N

Newton, Sir Isaac
  binomial expansion, 163–164
  divided differences, 84
  sum formulas, 102–103
Numbers
  complex, 35
  irrational, 29
  natural, 1
  transcendental, 29–30

O

Oldenberg, Henry, 164

P

Partial fractions, 87
Partitions
  of an integer, 165
  of a set, 143
Pascal's triangle, 53

Peano, Giuseppe, 1
Pell's equation, 62
Permutations
  cycles in, 152
  fixed points, 152
  inversions, 160–161
  left-to-right maxima, 160
Pigeonhole principle, 188–194
Plimpton 322, 60
Pólya, George, 106
Polynomials
  Bernoulli, 51
  Chebyshev, 95
  degree of, 74
  division algorithm for, 74
  factor theorem for, 74–75
  uniqueness theorem for, 74
Prime number theorem, 8
Principle
  inclusion-exclusion, 178
  mathematical induction, 11–12
  pigeonhole, 188
  well-ordering, 12
Product rule, 142, 163
Progressions and sums, 46–54

Q

Quadratic reciprocity, 65–71

R

Ramanujan, Srinivasa, 165
Rational root theorem, 30
Recurrence relations, 149–156
Remainder theorem, 74–75

S

Schwinger, Julian, 92
Spencer, Joel, 206
Sperner, Emanuel, 200
  theorem, 200–201
Stanley, Richard, 200
Stirling numbers
  first kind, 159
  second kind, 158
Sum rule, 142, 163
Sylvester, James, 167
Symmetric functions, 99–104
Synthetic division, 75

T

Tartaglia, 92
  method, 96
Telescoping sums, 52
Terquem's problem, 151
Tomonaga, Sinichiro, 92
Turán, Paul, 15
  theorem, 209

V

Vallée Poussin, C. J. de la, 8
Vandermonde, Alexandre
  determinant, 149
  identity, 177

W

Wiles, Andrew, 57